# A Brief History of Everything Wireless

Petri Launiainen

# A Brief History of Everything Wireless

## How Invisible Waves Have Changed the World

 Springer

Petri Launiainen
IT & Telecoms Professional
Brasília
Brazil

ISBN 978-3-319-78909-5    ISBN 978-3-319-78910-1   (eBook)
https://doi.org/10.1007/978-3-319-78910-1

Library of Congress Control Number: 2018937351

Printed on acid-free paper

This Springer imprint is published by the registered company Springer International Publishing AG
part of Springer Nature
The registered company address is: Gewerbestrasse 11, 6330 Cham, Switzerland

*To Kirsi, Titta and Taru*

# Preface

We humans are a curious bunch, constantly pushing the boundaries of science, attempting to find something new and interesting. The discoveries we make often seem to be of no immediate value, only to create billion-dollar industries just a couple of decades later.

Adapting to all these novel inventions has changed our lives relentlessly over the last hundred years, and not least by the unshackling effect of wireless communications: if we so choose, we can now remain in contact with the rest of the humanity wherever we roam, and have the unprecedented power of instant access to up-to-date information, using devices that comfortably fit in our pockets.

We are now constantly surrounded by technology that "just works", but have become totally oblivious to the enormous amounts of research and development that got us here: the underlying physics and processes vanish behind the magic curtain of engineering and friendly user interfaces.

The amalgamation of wireless communications and computing technology is changing our society faster and more widely than anything since the introduction of electricity into our cities and homes. As a result of this revolution, personal wireless connectivity is now more readily available than access to clean, running water.

Smartphones are everywhere, and many developing countries have been able to skip the installation of costly wired networks and jump directly into the world of untethered connectivity. This is mostly driven by the worldwide deployment of cellular networks, but also through the abundance of Internet-connected Wi-Fi networks, which have become an indispensable feature in our homes and public places.

At the other end of the scale, complex satellite technology now enables connectivity for the most remote locations in the world, from isolated islands of the Pacific Ocean to the North and South Poles.

The history of wireless communications is full of intriguing stories of personal triumphs and stinging defeats: it includes highly public clashes between individuals, corporations, and even nations. The outcomes of wars have been affected by the use of wireless technology, and the societies we live in today would be very different if the application of these invisible waves had been delayed by just a couple of years.

The quantity of available detail in this line of history is enormous, and to keep it all on a palatable level, I have chosen to compile a selected set of interesting storylines of prominent events, individuals and companies, explaining the necessary bits of the underlying technology along the way. Therefore, as the title says, this is a *brief* study of the phenomenal progress that has followed from the harnessing of the electromagnetic spectrum—many interesting stories and details simply had to be left out.

My focus is in revealing the direct and indirect consequences that the deployment of these new inventions has caused on society, from the early days of radio to modern cellular networks. I have also included some of the more esoteric and not so obvious uses of wireless technologies, in order to stay true to the word *everything* in the title.

Following the storyline does not require prior knowledge of the underlying technologies discussed—the essential bits are explained along the way. In case you want to delve a bit deeper on the technology side, the text contains references to *TechTalk* chapters that can be found at the end of this book. These are directed at readers who want to have more information about the "magic" behind it all.

This is not a study book—I deliberately decided against including any mathematical formulas, sticking to textual explanations instead. When presenting quantities, I chose to follow the metric system, leaving out conversions to imperial units, simply to maintain the flow of the narrative. The only exception is the use of feet for altitudes when discussing subjects of aviation, as this is still the global aviation standard, with the notable exceptions of China and North Korea.

The US dollar still remains as the reserve currency of the world, and hence it has been used for any monetary values presented in this book.

The further into the past you delve in, the hazier the details become: sometimes, I found that even key dates, unit counts, personal histories of key people, and various "history firsts" were quoted very differently by different sources. As with any book with multiple sources, the writer has to make a decision on what to accept as the most plausible description of events. I accept responsibility for any glaring errors that may have resulted in from this approach.

Researching for this book took two years, and I had many eye-opening revelations during the process. I sincerely hope that the following thirteen chapters provide you with new perspectives on the history of wireless technologies, and that you will find it to be a worthwhile and entertaining use of your time.

For the latest updates, comments, discussion, and links to interesting sources of information, please visit *http://bhoew.com*

Brasília, Brazil                                                                    Petri Launiainen
April 2018

# Contents

# Glossary of Units

Many of the following units are referred to across the text, so they are listed for your convenience here:

## Frequency Units

| | |
|---|---|
| Hz (hertz) | Oscillations per second |
| kHz (kilohertz) | Thousand (1000) oscillations per second |
| MHz (megahertz) | Million (1,000,000) oscillations per second |
| GHz (gigahertz) | Billion (1,000,000,000) oscillations per second |
| THz (terahertz) | Trillion (1,000,000,000,000) oscillations per second |

## Communications Speed Units

| | |
|---|---|
| bps | Bits per second |
| kbps (kilobits per second) | Thousand (1000) bits per second |
| Mbps (megabits per second) | Million (1,000,000) bits per second |
| Gbps (gigabits per second) | Billion (1,000,000,000) bits per second |
| Tbps (terabits per second) | Trillion (1,000,000,000,000) bits per second |

## Data Size Units

| | |
|---|---|
| bit | The smallest storage unit, either value "0" or "1" |
| byte | A group of eight bits |
| kB (kilobyte) | Thousand (1000) bytes |
| MB (megabyte) | Million (1,000,000) bytes |
| GB (gigabyte) | Billion (1,000,000,000) bytes |
| TB (terabyte) | Trillion (1,000,000,000,000) bytes |

# Chapter 1
# Tsushima Strait

On the night of May 26th, 1905, a convoy of thirty-eight Russian warships of the *Second Pacific Squadron* entered the Tsushima Strait. The ships were on transit from the Baltic Sea to the Russian city of Vladivostok, and their passage through the strait—the wide stretch of open water between Japan and the Korean Peninsula—was the last leg of an arduous journey across half the globe.

The trip that originated from the city of St. Petersburg had not started well: the paranoia amongst the Russians over the vastly overestimated reach of the opposing Japanese Navy during the ongoing Russo-Japanese War was so rampant that it had caused an inexplicable clash between the fleet and a set of unarmed British fishing trawlers that were mistaken as Japanese torpedo boats.

As a result of this bizarre shootout at Dogger Bank in the North Sea, Great Britain, which controlled the Suez Canal at the time, had revoked the right for the fleet to pass through the Canal. Therefore, the fleet had been on the move for eight months and had traveled a distance of 33,000 kilometers, going all the way around the Cape of Good Hope at the tip of the African continent. The ships were in dire need of comprehensive maintenance after such a long time at sea: their crews were exhausted and in low morale, and the hulls of the boats were heavily fouled by all the tiny sea creatures and plants that had attached themselves below the waterline, severely reducing their speed.

But the fleet had to keep pushing forward because the completion of this trip was of utmost importance—it offered the final chance for Russia to turn the tables in the ongoing war that had been going badly for them from almost the very first day.

Japan had started the war against Russia in February 1904 with a surprise attack against Port Arthur, Russia's naval outpost in China. Port Arthur, which is currently known as Dalian, had been the base for the naval forces of the *First Pacific Squadron*, and even though the actual damage suffered in the initial two days of the attack was relatively small, the situation soon started to deteriorate for the Russians.

In order to protect the bay of Port Arthur from further approaches by the Japanese naval forces, the minelayer *Yenisei* was sent to block the entrance of the harbor. Unfortunately, *Yenisei* hit one of its own mines in the process and sank,

© Springer International Publishing AG, part of Springer Nature 2018
P. Launiainen, *A Brief History of Everything Wireless*,
https://doi.org/10.1007/978-3-319-78910-1_1

losing 120 of its crew of 200, along with the map of the newly laid minefield. When another ship, *Boyarin*, was sent to investigate the situation, it also hit one of the newly laid mines. Despite frantic efforts to keep *Boyarin* afloat, it had to be abandoned, and while it was left drifting, it hit yet another of the mines laid by *Yenisei* and sank.

During the following months, several naval clashes occurred in the Yellow Sea as the Russians attempted to break out of the siege of the bay, and they did manage to cause considerable damage to the Japanese Fleet. But as the Russian losses continued to pile up through the summer months, the Japanese Navy eventually decimated the *First Pacific Squadron*, thanks to the effective command of Admiral Tōgō Heihachirō Saneyoshi.

The Japanese Navy was very lucky to have someone like Tōgō at the helm: he was a highly disciplined and skillful officer who, after having studied naval science in England, had a comprehensive international background—something that was quite unusual for a commander in the Orient at the beginning of the 20th century.

The new, supporting fleet of Russian ships that was entering the Tsushima Strait had initially been sent with the intention of joining the existing forces at Port Arthur and then pushing out the Japanese Navy. This maneuver would have kept access to Port Arthur open, allowing additional ground troops to be sent to the area, but, as Port Arthur was overrun while the fleet was still en route, the plan had to be readjusted accordingly: on new orders from St. Petersburg, the *Second Pacific Squadron* was directed to continue all the way to Vladivostok, resupply the ships there and then return to Port Arthur, engaging the Japanese Navy with fresh, hopefully definitive, force.

In order to reach Vladivostok as soon as possible, the fleet had chosen the shortest route that went past southwestern Japan along the Tsushima Strait, which even at its narrowest point was still about 60 kilometers wide, and therefore was expected to have plenty of room even for the large number of ships of the fleet to slip through.

The conditions at sea on that night in May were very promising—the weather was foggy and the Moon was in its last quarter, meaning that it would only rise after midnight.

The fleet pushed on, keeping a fair distance from normal shipping routes, trying to avoid any traffic in the area, including the Japanese scout ships that Admiral Tōgō had positioned around the strait. Tōgō was well aware of the approaching fleet and had assumed correctly that due to the rapidly deteriorating condition of the ships, the Russians would choose to take the shortest route towards Vladivostok via Tsushima Strait.

Despite almost optimal weather conditions, luck was not on the Russian side: in the early hours of the morning of the 27th of May, the Japanese cruiser *Shinano Maru* detected the navigation lights of the hospital ship *Oryol*, and after moving closer to investigate, noticed the shapes of multiple other ships in the convoy.

Although the position of the enemy was exposed relatively far from the land, the fate of the Russian fleet was sealed by the fact that *Shinano Maru* was equipped with an on-board radio transmitter—a Japanese copy of the newly introduced *Marconi* marine radio.

Thanks to this novel device, a short message stating *"enemy is in square 203"* was sent to the headquarters, informing admiral Tōgō about the exact location of the Russian convoy. Tōgō immediately ordered the Japanese Navy to intercept with all available vessels, sending a total of eighty-nine ships steaming towards the now exposed location of the Russian fleet. As a result, after almost two days of intensive fighting, the *Second Pacific Squadron* was utterly decimated: twenty-one Russian ships were sunk, eleven others were taken out of action, and over 4,000 Russian sailors were killed, while the Japanese side lost only about 100 sailors and three small torpedo boats.

Out of the original fleet, only one cruiser and two destroyers managed to break through the Japanese lines and eventually made their way to Vladivostok.

Thanks to this massive victory, the Russo-Japanese War was effectively over, giving Japan free rein as the undisputed military power in the region.

Russia, on the other hand, had lost its capability for any naval operations in the Far East and had only a handful of ships left in the Baltic Sea, barely able to protect St. Petersburg, as the ships of the *Second Pacific Squadron* were all originally part of the Russian *Baltic Fleet*.

The definitive result of this naval battle that ensued after that foggy night made headlines around the world and ended up having enormous geopolitical consequences: it tarnished Russia's reputation as a formidable international power, severely weakening the political clout of Emperor Nicholas II. This provided one more cause of disillusionment to the people of Russia, which in turn strengthened the forces pushing for a revolution. The sudden loss of Russian prestige also upset the existing balance of powers in Europe, which was one of the seeds that led to the First World War.

On the Japanese side, this undisputed and overwhelming victory provided the Japanese military command with a new sense of superiority, propelling Japan into an era of strong militarism. Tōgō became a legend, and is still highly appreciated, although he probably would not want the admiration of the hardline right-wing Japanese, who up to this day refuse to accept the facts of the atrocities that followed from the emboldened national sense of military and racial superiority that followed the Russo-Japanese War. The most notable of these disputed calamities is the *Nanking Massacre* of 1937–1938, which occurred during the Japanese-initiated Second Sino-Japanese War, and for which the casualty count is estimated to be between 50,000 and 300,000 civilians.

What further led to the deterioration of the overall political situation in Japan was the fact that the top ranks of the Japanese Military were lulled into a sense of invincibility after the many successful operations that followed the Battle of Tsushima. As a result, the commanders often acted without or against direct orders from the weak political leadership in Tokyo.

Eventually this kind of mindset would lead to an overreach, and it finally happened on December 7th, 1941: the Japanese performed a Port Arthur-style surprise attack again, this time against Pearl Harbor in Hawaii, thus dragging the formerly passive United States into the Second World War.

Although the Japanese were initially able to invade large parts of South-East Asia and the Pacific, the resources of the island nation soon became strained. This grave miscalculation over long-term sustainability of military actions eventually led to the unconditional Japanese surrender in 1945, days after the atomic bombs had decimated the cities of Hiroshima and Nagasaki.

As a side effect of forcing the United States into the war, the American troops also joined the Allied forces in Europe, providing massive material support for the Soviet Union as part of the fundamental shift in American war policy. This aid was essential in turning the tables against Germany on the Eastern Front.

It is fair to say that without the Japanese-induced American entry to the Second World War, an eventual German invasion of the British Isles, *Operation Sea Lion*, would have been a more probable endgame than *Operation Overlord* which eventually happened on the beaches of Normandy.

After the war, the Soviet Union quickly became a superpower with its own nuclear weapons and, in a few short years, polarized the world by turning into a wholesale exporter of Communism. The resulting Cold War isolated half of formerly united Europe behind the Iron Curtain and together with the developments in China, created several global conflicts that remain unsolved to this day, like the stark division of the Korean Peninsula.

On the Chinese side, the inability of the Chinese forces to stop the atrocities performed by the Japanese during the Manchurian Occupation weakened the position of the Chinese leader Chiang Kai-shek, giving rise to the communist movement led by Mao Zedong. Mao eventually took control over mainland China in 1949 and proceeded with his bold but misguided vision of Cultural Revolution, which ended up killing at least 30 million Chinese. Remarkably, the Chinese civilian casualties resulting from this internal political purge ended up being twice as numerous as those who had been killed by the Japanese between 1937 and 1945 in the Second Sino-Japanese War.

Many of these tremendous geopolitical events that completely altered the history of so many nations can be seen as direct or indirect consequences of that one fatal night at the Tsushima Strait.

Naturally it is impossible to predict what might have happened if the Russian fleet had made it safely to Vladivostok and successfully fought against Japan, but we can speculate:

Would the United States have remained true to their policy of impartiality during the Second World War, and as a result, would Japan have been spared the horrors of Hiroshima and Nagasaki?

And if Japan hadn't dragged the United States into war, would Nazi Germany have managed to overtake all of Europe, and eventually declared war against the United States?

Would this in turn have led to a nuclear war in Europe, as the *Manhattan Project* that created the nuclear weapons was principally a race against the competing developments in Nazi Germany, not Japan?

And locally in the Asia region, could China have developed peacefully under Chiang Kai-shek, preventing the rise of Mao Zedong and the formation of Communist China?

Would 50 million Chinese lives have been saved as a result? But without Deng Xiaoping as the successor for Mao, would China have remained the rural backwater it was at the beginning of the 20th century, and would we now have just one, unified Korea as a major global business power?

There are many alternative outcomes of history that may have panned out very differently if *Shinano Maru* had not had a radio transmitter on board.

That short message, sent via a technology that was barely five years old, had a profound effect on the history of the world.

One notable aspect to consider is the fact that wars have a strong tendency to speed up technological development, which eventually benefits the surviving post-war societies. Therefore, without *Shinano Maru's* short radio message and the events that it put in motion, we might only now be taking the first steps in computing or spaceflight.

The potential alternative ramifications are naturally quite impossible to conclude, but it is very clear that without these invisible waves and the numerous direct and indirect effects of their application on the course of history, the world would be a very different place.

So how did all this begin?

# Chapter 2
# "It's of No Use Whatsoever"

Inventors seem to come in two basic categories:

The first group includes those who are happy to tinker with something new and enticing, up to the point of proving a new concept or theory, only losing interest thereafter and moving on to the next problem.

The second group includes those who can perceive the potential value of the practical application of their inventions and, with the aim of extracting the maximum possible financial gain, strive to commercialize their findings as soon as possible.

It is pretty obvious that the German physicist Heinrich Hertz belonged to the first group.

Fourteen long years had passed since James Clerk Maxwell, a brilliant Scottish scientist, had published his groundbreaking theory of propagation of electromagnetic waves in 1873. His equations correctly predicted the existence of oscillating electric and magnetic fields, which, he calculated, would travel through empty space with a speed that closely matched the speed of light.

Many researchers had tried to prove that these mysterious, invisible waves really existed, but it took a brilliant German physicist to finally accomplish this feat. The astonishing part was that when Heinrich Hertz was queried about the possible uses of his newly made invention, he replied:

> It's of no use whatsoever ... this is just an experiment that proves Maestro Maxwell was right—we just have these mysterious electromagnetic waves that we cannot see with the naked eye. But they are there.

The obvious benefits of being able to transmit information and energy instantly with phenomenal speed and without direct physical connection seemed to be completely outside of his train of thought. He just wanted to prove a promising theory—nothing more.

Sadly, the life of this genius ended at the age of thirty-six, but his legacy lives on: *hertz (Hz)* is the unit for *frequency*—the number of oscillations per second—chosen by the *International Electrotechnical Commission (IEC)*, forty years after his death. Therefore, hertz is the fundamental unit used to describe anything from the extremely

© Springer International Publishing AG, part of Springer Nature 2018
P. Launiainen, *A Brief History of Everything Wireless*,
https://doi.org/10.1007/978-3-319-78910-1_2

low-frequency infrasound waves created by earthquakes to the high-energy bursts of gamma rays from collapsing stars in faraway galaxies.

And Heinrich Hertz rightfully deserves the credit.

It took another, totally opposite personality to eventually win the gauntlet of the moniker, "The Father of Radio": this title was earned by an Italian electrical engineer called Guglielmo Marconi, who spend large parts of his business career in Great Britain. He started working in this area in 1892, when he was just eighteen years old: his neighbor, Augusto Righi, who was a physicist at the University of Bologna, introduced Marconi to the research of Hertz. Marconi was fascinated by the idea of wireless communications and started his own research, trying to improve on Hertz's earlier experiments.

His breakthrough came when he realized that adding an *antenna* to his apparatus, both on the sending and the receiving side, was the key to expanding the range of communications. Marconi also noticed that the effect of *grounding* the system, or connecting the electric ground of the equipment to conductive soil, appeared to help in extending the range.

When you read the biography of Guglielmo Marconi, it depicts an able, productive, and, first and foremost, a highly opportunistic person. During his long career, Marconi did not worry about borrowing other inventors' ideas, as long as they improved the devices he was working on. Sometimes he ended up buying the associated patents later in the process, oftentimes not: when Marconi was awarded the shared Nobel Prize in Physics in 1909 with Karl Braun, he admitted that he had "borrowed" many of the features from the patents that Braun had filed.

Despite these sometimes questionable practices, Marconi was an excellent integrator of new ideas, and appeared to approach problems from a technical rather than a scientific perspective. His improvements came about mainly through relentless iteration, testing one slightly modified prototype after another, not necessarily understanding the underlying physics that were causing the improved functionality: he just kept on trying until he found something that worked. Creating a viable business based on his inventions was clearly his main objective—he strove to produce the best possible equipment and then moved on to aggressively market them, trying to extract the highest possible price.

Thanks to his numerous, often incremental patents, he seized the financial opportunities with gusto and proceeded with full-scale commercialization of this novel technology. In 1897 he set up a company, *Wireless Telegraph & Signal Co Ltd*, the initial funding of which was greatly helped by the fact that Marconi came from a very rich family: his mother, Annie Jameşon, was the granddaughter of the founder of the *Jameson Irish Whiskey Company*.

Marconi originally tried to get the Italian government interested in his research, but did not manage to raise any interest, so he explained his *wireless telegraph* idea to the *British Postal Office* through the contacts that his mother had provided, and finally got the positive feedback he needed to go forward with his business interests.

As often happens, success abroad also brings success at home: after convincing the British that his invention was useful, the Italian Navy bought radios from

Marconi, and even made him responsible for the Italian military's radio service during the First World War.

At this time, his reputation was already well established, both through his business activities, and due to the fact that he had been awarded the shared Nobel Prize in Physics in 1909.

Marconi kept a relentless, sharp focus on radio technology throughout his entire life, changing the company name to *Marconi's Wireless Telegraph Company Ltd* in 1901, and the name "Marconi" became the synonym for radio in the early decades of the 20th century.

When Marconi died at the age of sixty-three, all official radio transmitters on the British Isles, together with numerous other transmitters abroad, honored him by being quiet for two minutes on the day of his burial.

A third very notable early radio pioneer was Nikola Tesla:

Tesla appeared to enjoy maintaining an image of a "Man of Wonders", often announcing his inventions with a big splash. He had a touch of the "Touring Magician" in him, and understood the value of putting together a good show with some hyperbole thrown in, just to advance his cause. He was clearly extremely intelligent, which explains his ability to bounce so effortlessly from subject to subject.

This inclination to constantly change his focus put him in the same inventor category as Heinrich Hertz, with an unfortunate side effect: even though he advanced many, very different technologies through his numerous inventions, he never really worked on reaping substantial financial benefits from his fundamentally genial solutions. Some of his patents did generate considerable sums of money, but he spent it all on the prototype of yet another interesting and grandiose concept. Other patents he relinquished in an almost nonchalant way, letting his business partners benefit from them, sometimes very generously indeed.

Tesla's various research areas of interest spanned from power generation to radio to wireless transmission of energy. He earned his reputation in the late 19th century by first working for another prominent inventor of the era, Thomas Edison, greatly improving Edison's *Direct Current (DC)* generator technology. According to some historical notes, Tesla was conned out of getting a promised financial reward for his achievements by no other than Thomas Edison himself, and he resigned in anger, starting a life-long feud against Edison. At the same time, another successful industrialist, George Westinghouse, had heard of experiments using *Alternating Current (AC)* in Europe, and knowing that Tesla was familiar with the subject, Westinghouse hired him to focus on improving this novel technology.

Tesla's designs proved AC technology to be superior to Edison's established DC technology, suddenly causing a direct financial threat to Tesla's former employer, who had been building a major business around the generation and distribution of Direct Current. As a result, Edison put his "Genius Inventor" reputation on the line, fighting against indisputable technical facts by all means possible. This epoch of bitter and very public fighting over the two competing technologies was aptly named *The War on Currents*. Edison even went as far as publicly demonstrating and filming the electrocution of several animals, including an elephant, with the use of Alternating Current, trying to prove the "inherent dangers of this inferior technology".

But the poor elephant and numerous other animals died in vain: Tesla's AC power technology, thanks to the ease of transferring it over long distances and the ability to adapt the voltage to local requirements through transformers, has survived to this day as the primary means of distributing electric power to users.

Despite this unprecedented victory, but very much in line with his character as a scientist and not a businessman, Tesla gave up his rights to the royalties gained from the use of AC generators utilizing his patents. He was driven to this by George Westinghouse, who was facing temporary financial difficulties, and managed to talk Tesla out of his existing, very lucrative agreement. Just a couple of years later, the patents relinquished by Tesla helped Westinghouse to regain his financial footing and he became very wealthy indeed.

Many of Tesla's later works were never properly documented, and a lot of his early research notes and prototypes were lost in a fire in 1895. His surviving notes are full of extraordinary claims and promises, but lack enough detail on how to achieve them. Thanks to this large amount of ambiguity, Tesla's potential inventions seem to be an endless source of various conspiracy theories even today.

The fact that Tesla's complex research needed a lot of funding makes it almost incomprehensible to understand his relaxed approach to money issues. At one time in his life, he even had to resort to working as a ditch digger in New York to support himself. When he managed to link up with new investors or gained some money from his consultancy work, he quickly spent his funds on his new experiments, and without short-term tangible results, the investors stopped funding him further. Many potential lines of research were cut off due to the lack of funds to complete them. Tesla's constant change of focus caused him to work in parallel with multiple ideas, which resulted in various prototypes with poor documentation.

All in all, Tesla was almost a total opposite to Marconi, who continuously improved and fine-tuned his products, with clear financial and technical focus in mind, borrowing from other inventors and trying to create monopolies for his businesses.

These two geniuses came head to head in one of the more controversial patent disputes of these early days, when in 1904, the U.S. Patent Office decided to revoke Nikola Tesla's patents for radio tuning circuitry and ratify the patents from Marconi instead. Tesla fought for years to reverse this decision, and whilst the U.S. Patent Office re-established some of Tesla's patent claims in 1943 right after his death, he never achieved his lifelong goal of being declared the man who had made the first ever radio transmission. Marconi was able to hang on to this claim with a solidly written patent from 1896 that includes the words:

> I believe that I am the first to discover and use any practical means for effective telegraphic transmission and intelligible reception of signals produced by artificially-formed Hertz oscillations.

Tesla had also been working on radio just before the turn of the century, and he had even demonstrated a radio-controlled model boat, the *teleautomaton*, at

Madison Square Garden in 1898. The demonstration worked so well that Tesla had to open the model to show that there really was no trained monkey hiding inside. Even though remote control like this could obviously have several military applications, Tesla failed to raise interest of the U.S. Army.

Another example of unfortunate timing appears to be when Tesla managed to take an X-ray image a couple of weeks before the existence of X-rays was announced by Wilhelm Röntgen. Therefore, although Tesla was pioneering the use of the extreme end of the electromagnetic spectrum in 1895, we refer to pictures created by X-rays as "Röntgen images".

In cases like this, along with many others that he consequently worked on, Tesla was clearly ahead of his time—apart from his groundbreaking work in the generation and transmission of Alternating Current, for which he had relinquished all his rights to George Westinghouse, the world was not ready for his inventions.

When Tesla did not manage to get traction with his radio controlled *teleautomaton*, he soon shifted his interest to investigating wireless transmission of power instead of wireless communications, providing some stunning demonstrations in his laboratories at Colorado Springs and Long Island. These demonstrations were not only huge but also very expensive, seriously draining even the ample funds that Tesla had been able to raise from various investors at the time. Tesla kept on pitching his ideas to his acquaintances in the energy business, but getting external funding for his ambition of providing "limitless, wireless free energy" did not bear much fruit—who would fund research for something that would potentially end up killing their existing, hugely profitable business model?

This shift in Tesla's interests left the advancement of wireless communications to Marconi and his competitors. Tesla willfully abandoned a market that was about to experience tremendous growth, making many of the other participants in this business very wealthy in the process.

In Marconi's case, his situation was strengthened by the dual role as both the patent owner and the owner of a company that produced functional devices based on those particular patents. With his name on the equipment placards and in headlines regarding new distance records, there was no doubt who was behind these new, almost magical devices. Marconi kept on demonstrating the continuous improvements that he managed to incorporate into his equipment. His worldwide reputation grew in sync with every new achievement and business deal.

In comparison, Tesla had managed to create flashy demonstrations and lots of hype in the media, but eventually it did not generate enough cash flow to cover his overly grandiose ideas.

Tesla was broke.

In 1934, George Westinghouse, having now enriched himself with the original inventions of Tesla, recognized his former employee's dire situation and started paying Tesla a small monthly "consultancy fee" (about 2,200 dollars in current terms), covering also the cost of his room at Hotel New Yorker.

To make things even worse, Tesla was hit by a taxicab only three years later, and his health never recovered. He died alone in his hotel room in January 1943 at the age of eighty-six.

All in all, Tesla's various accomplishments throughout his lifetime are very impressive, and in many of his designs, he was far ahead of his time. He also made this eerily accurate prediction in 1926:

> When wireless is perfectly applied the whole earth will be converted into a huge brain, which in fact it is, all things being particles of a real and rhythmic whole.
>
> We shall be able to communicate with one another instantly, irrespective of distance.
>
> Not only this, but through television and telephony we shall see and hear one another as perfectly as though we were face to face, despite intervening distances of thousands of miles;
>
> and the instruments through which we shall be able to do his will be amazingly simple compared with our present telephone. A man will be able to carry one in his vest pocket.

If you make a link between "the whole earth converted into a huge brain" and the Internet of today, Tesla not only foresaw the wireless communications revolution, but also the computer revolution that has led us now to be able to check any detail, anywhere, using our smartphones.

But Tesla could not predict the explosion of conspiracy theory websites that have sprouted all around this modern version of Tesla's "huge brain", caused by Tesla's almost mythical reputation: his versatility and the fact that he made numerous boisterous claims without solid proof has left him with an aura of mystery, the perceived strength of which is too often directly relative to the thickness of the reader's tin foil hat. The number of books and videos that push the more esoteric and poorly documented research of Tesla is astounding, most of them naturally claiming that "the evil government" is keeping us from the limitless benefits that Tesla's "forcibly hidden inventions" would bring to the mankind.

What probably greatly helped to initiate these rumors was the fact that the *Federal Bureau of Investigation (FBI)* seized Tesla's belongings two days after his death, treating him as if he was a foreign citizen. There was no real reason for this, as Tesla, although originally an immigrant from Serbia, had become a naturalized American citizen already in 1891.

The resulting report of a study of Tesla's final belongings found nothing of interest, stating that:

> [Tesla's] thoughts and efforts during at least the past 15 years were primarily of a speculative, philosophical, and somewhat promotional character often concerned with the production and wireless transmission of power; but did not include new, sound, workable principles or methods for realizing such results.

But naturally, a "nothing to see here"-style report from the FBI for a conspiracy theorist is like a red flag for a bull...

Tesla finally got his name in the "Technology Hall of Fame" in 1956, when the *International Electrotechnical Commission (IEC)* honored his achievements by choosing *tesla (T)* as the unit for *magnetic flux intensity* in the *International System of Units (SI)*. Hence Nikola Tesla, along with Henrich Hertz, will be remembered for all eternity.

Parallel to Tesla's ever-changing research focus, Marconi's strict focus on radio technology kept on producing new achievements in terms of the distance over which his equipment was able to make contact: he became the first to achieve wireless transmissions over the Atlantic Ocean in 1901. Although the demonstration only consisted of the reception of a single Morse code letter "s", it proved the huge potential of radio technology: for the first time in history, the Old and the New Continents were joined without wires in real-time.

Morse code and various radio technologies, including the early spark-gap transmitters, are explained in TechTalk *Sparks and Waves*.

At the time of this milestone by Marconi, intercontinental communications had been possible for over 30 years, but only through cables at the bottom of the ocean. The first Transatlantic Telegraph Cable had been in place since 1866, allowing Morse code transmissions at a speed of a whopping eight words per minute due to the considerable signal deterioration across such a long distance. At the turn of the century, only a handful of similar cables existed, and laying down a new one was vastly expensive compared with the installation of radio equipment on both shores of the Atlantic Ocean.

Marconi's achievements made front-page news, and he was well on his way to becoming a worldwide celebrity. He immediately took full advantage of his fame and the lax corporate legislation of the early 20th century, trying to monopolize his position in wireless communications before his competitors could catch up.

For example, one of the key requirements of his licensing deal with *Lloyd's Insurance Agency* in 1901 was that communications between Marconi and non-Marconi equipment were not allowed. Technically this was a purely artificial limitation that went against the grain of most of his customers, but was accepted, thanks to the high quality of Marconi's equipment at the time.

Realizing that the most obvious market for his equipment was in sea-to-shore communications, Marconi attempted to consolidate his situation by installing shore-based radios at major ports along the shipping lanes. In this way he gained a solid foothold for his equipment in the most prominent and critical locations.

Exclusive deals like this helped Marconi expand his early sales, but the demand for radio equipment was so huge that the potential for large profits soon started pulling several other manufacturers in the game.

The basic principles of radio were reasonably simple for engineers to understand and copy, and eventually some manufacturers were even able to surpass the quality of Marconi's own products. As a result, despite being continuously able to improve his technology, Marconi's demands for a monopoly were starting to sound increasingly hollow: it was obvious to all technically aware customers that radio waves were a shared commodity, and the name that was on the device had nothing to do with the fundamental physics of transmitting and receiving signals.

Something was bound to go awry for a company insisting on a monopolistic position for a system that, by default, had inherent interoperability, and for Marconi, this happened in early 1902:

Prince Heinrich of Prussia, the brother of Kaiser Wilhelm II, tried to transmit a "thank you" message to President Theodore Roosevelt while he was returning from a trip to the United States on board of SS *Deutschland*. But due to the fact that the German ship was using a German *Slaby-Arco AEG* radio, the Marconi-equipped shore station at Nantucket refused to relay the message that was otherwise perfectly received by the "non-compatible" Marconi equipment.

The technologically literate and now very irate Prince contacted his elder brother the Kaiser after the trip, urging him to set up a global meeting over the control of radio communications. This incident led to the first international conference on the issue just one year later.

Although the American delegates had initially been lukewarm to the prospect of enhanced governmental control and the demand for cross-manufacturer collaboration, such regulations came into effect in 1904. They mandated governmental control over all shore radios and further dictated that the military would take absolute control over any radio facilities in times of war.

Marconi started a furious disinformation campaign to fight against this decision: when the Navy removed Marconi's equipment from the Nantucket Shoals Lightship, the representatives from Marconi's company claimed that not having their equipment at both ends of the wireless communications would seriously endanger the connectivity between ships and the shore station. The Navy, however, through their internal testing efforts, knew that cross-manufacturer interconnectivity was fully possible, and the *Bureau of Equipment's* comment on Marconi's claims was blunt:

> [This is an] audacious attempt to induce the Government to participate in a monopoly calculated to extinguish other systems, some of which are more promising than the Marconi.

With this move, the U.S. Navy changed the coastal wireless services from the earlier, artificial monopoly to an "open-to-all" operation, where ship-to-shore service was free of charge.

Very soon this service grew to include useful extensions like time signals from shore to ships to aid in navigation, and to relaying of weather reports from ships to shore. On the 26th of August 1906, SS *Carthago* sent the first hurricane warning off the coast of Yucatan Peninsula, and soon ships passing certain areas of the Atlantic Ocean started sending regular weather reports.

Thanks to this up-to-date information, it become possible for other ships to alter their course accordingly, thus avoiding the areas of worst weather. Starting in 1907, radio warnings of "obstructions dangerous to navigation" were added to daily transmissions, including information on lightships that had veered off their designated location, as well as notifications of lighthouses with lights out of commission. These relatively simple applications of radio technology improved the seafarers' safety immensely, truly showing the potential benefits of wireless real-time communications.

Due to the improved safety that all these real-time messages provided, radio equipment was on its way to becoming an essential part of all ships that were large enough to support the necessary, massive antenna structure.

Still, radios were very expensive, and mostly used on an on-demand basis only. For passenger ships, a large part of the usage was to act as a medium of vanity for passengers wanting to brag about their travels by means of off-shore messages relayed to their friends: for the high society, it was "hip" to send a telegram that originated off-shore.

Compare this with the millions and millions of *Facebook* and *Instagram* updates emanating from holiday destinations today.

The fact that radio was still seen as an optional extra had some serious side effects: an alarming example of this occurred in relation to the fatal accident that happened to RMS *Titanic* in the early hours of April 15th, 1912. Although RMS *Titanic* had brand-new *Marconi* radio equipment on board, the distress signals it sent were not heard by the nearest ship, SS *Californian*, simply because the radio operator on SS *Californian* had ended his shift only ten minutes before the accident.

Had SS *Californian* received the message, hundreds more of the passengers on the doomed RMS *Titanic* would probably have survived.

An even worse coincidence was that several iceberg warnings received by the radio operator on board RMS *Titanic* never left the radio room, as the operator was busy continuously relaying personal messages from the passengers instead. Technical problems with the transmitter the day before the accident had caused a huge backlog of these "important" passenger messages to pile up, and the radio operator had his hands full.

On top of this procedural failure to sensibly prioritize the communications, the budding technology itself still had some serious issues. For example, the poor selectivity of all current electro-mechanical radios caused interference between adjacent channels. Hence, it had been impossible to get a clear reception of some of the iceberg warnings that were transmitted by other ships during RMS *Titanic*'s doomed Atlantic crossing.

Despite all these limitations, the existence of radio on board of RMS *Titanic* was crucial in terms of the rescue effort that followed. It is quite likely that in the weather conditions that prevailed on that night, without the crucial radio call, not a single passenger would have been rescued alive, and the disappearance of an "unsinkable" ocean liner on its maiden voyage would have become one of the unresolved mysteries of the century.

As a result of what happened to RMS *Titanic*, especially in relation to the lack of a round-the-clock listening watch on SS *Californian*, the United States Senate passed the *Radio Act of 1912*, mandating all ships to have 24-hour monitoring of distress frequencies.

Another early case that demonstrates the usefulness of radio during emergencies happened six years before the RMS *Titanic* incident, in the aftermath of the violent 1906 earthquake in San Francisco. The tremors that leveled the city also knocked out all local telegraph stations and the entire telephone network, leaving the city in a complete communications blackout. The only working telegraph connection after the earthquake was on nearby Mare Island, which also had radio equipment on site. Steamship USS *Chicago*, which had just sailed off San Francisco prior to the earthquake, heard the news via its on-board radio, immediately returning to the shore and starting to act as a relaying station between San Francisco harbor and

Mare Island. With this makeshift wireless link, San Francisco was again connected to the rest of the United States.

With multiple similar occurrences, the benefits of wireless connectivity were duly noted.

Thanks to the earlier removal of the artificial *Marconi* monopoly and the free competition that followed, both the development of this new technology and the scale in which it was being deployed expanded rapidly.

And where there is demand and hype, there are always opportunists trying to swindle gullible persons out of their savings. Some persons utilized the media buzz around radio's great promise, together with weak patent claims, purely as a way of propping up the stock value of their companies. They lured hapless investors with comments like:

A few hundred dollars invested now will make you independent for life.

As a result, the world witnessed its first wireless technology bubble, with companies that quickly went bust after succeeding in selling their questionable stock to hopeful outsiders.

One name to mention, as both good and bad, is Lee de Forest, whose *American De Forest Wireless Telegraph Company* witnessed exactly the kind of boom-bust cycle presented above, although admittedly de Forest can't be entirely blamed for the demise of the company that carried his name.

De Forest himself was a fierce litigator against other inventors, and he spent a small fortune while pushing his claims, some of which did not hold water in court. In contrast, he also ended up being sued for deceit by no less than the United States Attorney General, but was later acquitted.

Despite his somewhat colorful business dealings, he made it solidly into history books as an inventor of *grid audion*, the first three-electrode *vacuum tube*.

Vacuum tubes enabled the amplification of weak signals and the creation of high-quality, high-frequency oscillators, hence revolutionizing both the receiver and the transmitter technologies. This crucial device catalyzed the *solid-state electronics* era that was further improved by the invention of *transistors* and later *microchips*, all of which are discussed in greater detail in TechTalk **Sparks and Waves**.

Grid audion is an excellent example of how some inventors take the work of their predecessors and improve it beyond its original purpose: as part of his attempts to create a durable filament for a light bulb, Thomas Edison had noticed a peculiar property of current flow in the vacuum bulbs he used for his experiments. He did not quite understand it, but he recorded what he had discovered and called it the *Edison Effect*: a way to restrict the flow of electrons in a single direction only.

Based on this effect, Sir John Ambrose Fleming, who had been one of Marconi's technical advisers, created the first two-electrode solid-state *diode*, or *thermionic valve* as it was called at the time. With this component it was possible to turn Alternating Current into Direct Current—something that was extremely hard to do utilizing only electro-mechanical technology. Although the thermionic valve was somewhat unreliable and required a high operating voltage, it further strengthened the position of Alternating Current as the means of distributing electricity.

In 1906, Lee de Forest then added a third electrode to the thermionic valve, thus creating the first grid audion: the predecessor of all vacuum tubes. Furthermore, in 1914, the amplification potential of vacuum tubes was improved by a Finnish inventor, Eric Tigerstedt, who rearranged the electrodes inside the tube into a cylindrical format, resulting in a stronger electron flow and more linear electrical characteristics.

With no moving parts, vacuum tubes were much more reliable and smaller than their electro-mechanical predecessors. They not only revolutionized radio technology but also started the era of versatile electronic systems in general, eventually leading to the very first computers.

In hindsight, if we look at the electronics revolution that was initiated by the grid audion, it is clear that Lee de Forest should have received a Nobel Prize for his invention. The grid audion was a true game changer, just like the transistor would be some 40 years later. But while John Bardeen, Walter Brattain and William Shockley got a Nobel Prize for their work on the transistor at the legendary *Bell Labs*, Lee de Forest unfortunately was left without one.

Finally, the *Radio Act of 1912* that was put in place after the RMS *Titanic* incident also cleared another, potentially problematic issue: the emergence of radio had drawn in a lot of technology-oriented enthusiasts to experiment with this new technology, and the existence of *Amateur Radio* activities was officially noted in the Radio Act for the first time. To keep these experimenters from interfering with officially mandated wireless communications, Amateur Radio was forced to utilize frequencies that were 1,500 kHz or above. These higher frequencies were thought to be useless at the time, which was soon proven to be a completely false assumption. Therefore, as technology and understanding of the behavior of radio waves has advanced, the frequencies that are designated for Amateur Radio use have been reallocated through the years.

On many occasions, the assistance of Radio Amateurs has had significant value: examples of this can be found in Chapter 3: **Radio at War**.

Radio Amateur activities remain a small but vibrant niche of radio experimentation and development, with a couple of million active Radio Amateurs around the world. The various classes of this activity range from local to global communications, even utilizing dedicated *Orbiting Satellite Carrying Amateur Radio (OSCAR)* relay satellites for bouncing signals across the globe.

# Chapter 3
# Radio at War

As soon as Marconi's marine radios became available, the Navies of the world saw the potential of being able to transmit messages to and from ships on the open seas.

The U.S. Navy did excessive tests using Marconi's equipment in 1899 and concluded that this technology should be deployed across the fleet. Additional pressure came from a report sent to the Secretary of Navy in December of the same year, which noted that the British and Italian Navies were already utilizing Marconi's radio equipment on their ships, whereas French and Russian Navies had radios from the French *Ducretet Company*.

The scramble of the armed forces to match the capabilities of their potential opponents was as rampant in the early 20th century as it is today, and, like it or not, military applications have always been one of the driving forces behind progressive leaps in technology. The area of wireless communication is no exception to this.

At the turn of the century, even though the namesake company of Guglielmo Marconi was the trailblazer for the available commercial radio equipment, the negotiations between the U.S. Navy and Marconi did not end well. This was all due to Marconi's highly opportunistic push: when the Navy asked for an estimate for twenty radio sets, Marconi wanted to lease them instead of selling them, asking for $20,000 for the first year and $10,000 for every year thereafter. In current terms, these sums would be about $500,000 and $250,000 per radio—a tidy flow of guaranteed income for Marconi's expanding business.

Despite the urge to get radios installed on their ships, the Navy was not willing to enter a leasing contract of this size with Marconi, deciding on a wait-and-see-approach instead. This short-term greediness by Marconi opened the first cracks in his grand plan to create a monopoly: with Marconi's Navy deal on hold, other manufacturers seized the opportunity and offered their solutions without any requirements for annual license fees. Hence, after excessive tests with equipment from various manufacturers, the U.S. Navy ended up buying twenty German *Slaby-Arco AEG* radios, which not only were considerably cheaper than Marconi's equipment, but according to the Navy's internal testing, also offered superior technology in terms of low interference and high selectivity.

© Springer International Publishing AG, part of Springer Nature 2018
P. Launiainen, *A Brief History of Everything Wireless*,
https://doi.org/10.1007/978-3-319-78910-1_3

This new competition shows the simplicity of the radio technology at the beginning of the 20th century: Adolf Slaby, a German professor with good connections to Kaiser Wilhelm II, and his then assistant, Georg von Arco, had participated in Marconi's experiments over the English Channel in 1897. Slaby understood the huge potential of this technology, especially for the military, and made a successful copy of Marconi's equipment, inducing the birth of the German radio industry, which soon grew as a formidable competitor for Marconi on a world scale.

The purchase of German-made equipment for military purposes sparked an immediate attack from another company of the time, the *National Electric Signaling Company (NESCO)*. Their representatives tried in vain to get the Navy to either purchase *NESCO's* "domestic" equipment, or at least start paying royalties on the *Slaby-Arco AEG* radios, which they claimed to have violated *NESCO's* patents. This forced the Navy to initiate an internal investigation into the referred patents, eventually concluding that the claim was without merit, and hence the purchase from *Slaby-Arco AEG* could proceed.

When the *Slaby-Arco* deal was announced, Marconi found himself effectively bypassed in a potentially enormous, long-term business opportunity. Realizing the implications that would result from losing the Navy's order, both financially and in terms of publicity, Marconi made another, license-free offer to the Navy. But trusting their own research results regarding the quality of the available equipment of the time, the Navy held their ground and went on with *Slaby-Arco AEG* equipment for their initial fleet deployment.

The deterioration of US–German relations a couple of years later forced the Navy to seek an allied supplier again, but there was no going back to the leasing approach of Marconi any more—the devices were always going to be bought, not leased.

Like in any good procurement process, the Navy forced the manufacturers to compete based on technical merits, adhering to the strict requirements and specifications set up by the Navy. This forced the companies to innovate, and considerable improvements both in distance covered and selectivity against interference were gained in a relatively short time:

On December 19th, 1905, a new distance record of almost 3,500 kilometers was made between the Navy's stations at Manhattan Beach, Coney Island, and Colon, Panama. This distance was longer than the length of the Atlantic undersea telephone cable between the west coast of Ireland and Newfoundland, and it enabled a connection to the construction site of the Panama Canal, without the huge cost of laying a cable on the seabed.

Although these long-distance wireless connections offered a novel, low-cost channel for communicating with remote naval bases, it was the naval strategy that was most affected by this new technology: up to this point, ships had only been able to communicate in line-of-sight conditions, using either signaling semaphores, flags or Morse code via signaling lights. While out on the open sea, sea-to-land communications was possible only by using homing pigeons, with all the obvious limitations of such a rudimentary and unidirectional solution. For any new orders to be received by the ships, they had to enter a port to use a telegraph, or at least be close enough to a shore to be able to use signaling lights.

The emergence of radio changed all this, as the ships could receive new orders at any time, as well as report their status, location and crucial observations whenever necessary. The tremendous power of instant communications could literally dictate the difference between life and death, and nowhere else would that be as essential as at a time of war.

Despite this new, instant communications channel, some commanders initially resented the prospect of receiving instructions while out on the high seas—they perceived that this new line of command diminished their own authority in making decisions. Suddenly they were no longer the undisputed commanders of their ships while seaborne, since it was now possible to receive new instructions from the Admiralty at any time.

A claim was also made that radio would be useless due to interference during real battles, and that any instructions would instantly be available to the enemy: after all, radio transmissions were free for anyone to listen. But the devastating consequences that followed from the use of the radio on board *Shinano Maru*, as discussed in Chapter 1: **Tsushima Strait**, completely annihilated any resistance towards this new technology.

The eavesdropping issue could be circumvented by the use of shared code books for message encryption, and such an approach had already been widely used throughout history. Therefore, the worst side effect of an enemy hearing the coded transmission would be the possibility of determining the relative distance of the transmitter, based on the strength of the received signal.

In terms of equipping battleships with radios, there simply was no going back after the battle of Tsushima Strait.

At the turn of the century, initial steps to utilize radio were also being taken amongst land-based troops, and the first such case happened in 1899 during the Boer War in South Africa. This was also the very first time that radio was used in an actual war situation.

The *Marconi* equipment that was sent to British troops in order to be used in naval installations was modified locally for land-based use, even to the extent that some parts from German radios made by *Siemens* were mixed with the original *Marconi* hardware. The source of *Siemens* radios was a shipment that had been intercepted by the British troops along the way, originally destined for the Boer forces.

The initial results were not too impressive, though:

A major problem was the massive antenna support rig setup, which, due to the lack of ship masts to hang the antenna on, had to be improvised from bamboo. If the wind was strong enough, large kites were used to lift up the antenna wire.

The local climate created another obstacle: interference from frequent thunderstorms in the area caused the rudimentary mechanical signal detectors, *coherers*, to jam, thereby rendering the receivers useless.

You can read more about coherers in TechTalk **Sparks and Waves**.

Additionally, the importance of grounding radio equipment for optimal transmission and reception was not yet well understood, limiting the achievable range even in cases when the rest of the equipment was fully functional.

Grounding was not an issue on ships, as salt water has an excellent conductivity and therefore was optimal for radio equipment grounding purposes. Therefore, one of the *Marconi* radios that was sent to South Africa was reinstalled on a ship anchored in the harbor and then used for communications, providing better coverage than land-based radios.

Some ten years later, during the First World War, the technology for portable radios for land-based use advanced rapidly. The jam-prone electro-mechanical coherer had recently been replaced by a signal detector based on *carborundum crystal*, removing the earlier jamming issue in the presence of frequent thunderstorms in tropical locations. More importantly, this solution was much more robust against the mechanical strains of being constantly hauled around. This greatly improved the reliability of the receiver in field conditions.

One of the early land-based systems was the *Knapsack Station*. With the required antennas, grounding mats and batteries, it weighed about 40 kg and hence required a four-man team to move it around. Although the term "portable" was still very different from what we expect from such equipment today, a movable station made a huge difference during land-based warfare with rapidly shifting front lines. With only a four-man team, it was now possible to set up a radio station at short notice and on demand at almost any location. Commands could be instantly dispatched and received, and the leadership was able to receive up-to-date information regarding the situation on the battlefield.

Even though the *Knapsack Station* still required a four-man crew, it was still a major improvement over the previous generation *Telefunken* product—a radio that was installed on a mule-pulled carriage, as well as the horse-drawn radio wagons that had housed the *Marconi* radios in the Boer War.

The first Radio Car also became available during the final parts of the First World War—an example of the benefits of combining two separate and rapidly developing technologies.

It was now clear that both enabling own communications and disrupting the enemy's communications capabilities were crucial parts of modern warfare. Hence, when the First World War had started, the British Navy immediately sprang into action: the underwater cables connecting Germany to South America and the United States were dredged up and cut, and German long-range wireless stations were attacked around the world.

These long-range radios were still a rarity and hence in high demand, and their potential value for the enemy was well understood: for example, in 1914, when the Germans were approaching Brussels, the Belgians decided to blow up a Belgian radio station that had a range of about 6,000 kilometers—leaving it intact would have helped the Germans to maintain contact with their troops in faraway battlefields.

Radio technology also expanded into aviation, with the new aspect of not using it only for communications, but also for navigation: the first radio-based navigation system was devised to help the German airships, the *Zeppelins*, determine their position while flying during the night or above the clouds, with no visible landmarks in sight. This novel navigation system made it possible to take advantage of adverse weather conditions, which made it harder for the enemy to detect and subsequently shoot down the slowly moving *Zeppelins* that otherwise would have been easy targets.

The accuracy was not sufficient, though, as the only indication that could be attained was the bearing to the station: distance information had to be calculated by other means. Therefore, in 1918, when the *Zeppelins* attempted to bomb London by radio navigation only, none of the bombs fell on land.

Improved versions of this initial system found their way into civilian use after the war, and although satellite-based navigation is currently becoming the norm, traditional land-based radio navigation aids are still the basis of commercial air traffic today, as will be discussed in Chapter 6: *Highways in the Sky*.

The newly invented vacuum tube technology was quickly adapted to military use, resulting in scaled-down equipment size and improved reception range due to better selectivity. The use of vacuum tubes made it possible to switch to use higher frequencies, and as the required antenna size is inversely proportional to the frequency used, higher frequencies required much smaller and therefore much more portable antennas. A large, erected antenna was easy to detect and often drew fire from the enemy the moment it was set up on the battlefield, so being able to quickly assemble and disassemble a smaller, less prominent antenna was a major improvement.

Last but not least, the vacuum tube technology enabled the use of speech instead of Morse code for communications, considerably speeding up information exchange: a skilled radio operator can transmit and receive 30–40 words per minute via Morse code, while spoken word can easily reach between 100 and 150 words per minute. Most importantly, by using voice communications, anyone who has basic training on how to operate the radio equipment per se, can step in and relay messages, without any knowledge of dots and dashes.

On the eve of the Second World War, portable radio equipment was an indispensable and essential part of any self-respecting army. Vacuum tubes had now become the norm, and compact radio equipment was installed in cars, tanks, airplanes and even small boats. Instead of having separate units for the receiver and transmitter, the concept of a combined *transceiver* became the norm.

Radio had already been used in tanks during the First World War, but the size of the antenna required the tank to stop, set up the equipment outside and establish a stationary radio station for the duration of the communications, leaving the tank vulnerable to attacks. But when Germany started the Second World War by invading Poland, the *Panzer* tanks had radios that could be used during combat without any stops. The unprecedented flexibility offered by constant radio access was a key component in the *Blitzkrieg*—the quickly advancing, technologically superior attack doctrine of Nazi Germany.

To maximize interoperability on the battlefield, the Allied forces worked hard on standardizing their communications equipment. Eventually this led to the adoption of one of the most inventive hand-held radios in the Second World War, the *BC-*

*611*. It was originally built by a company called *Galvin Manufacturing*, which had been producing car radios prior to the war. Due to the huge wartime demand, it ended up being manufactured by several other companies in the Allied forces' countries. By the end of the war, over 130,000 *BC-611* units had been produced.

The *BC-611* was the first real *Walkie-Talkie*, a truly portable transceiver that had all parts integrated into a single 2.3 kg unit, including space for two sets of batteries: one for heating the vacuum tube filaments and another for the actual communications circuitry. The size of the unit was such that it was easy to operate with one hand. Having a complete transceiver in such a small, compact and water-resistant unit was a technological marvel at the time, considering that it was still based on vacuum tube technology. Perhaps most importantly, it had a very simple user interface: to turn it on, you extended its antenna, and then simply pushed a button to speak. The frequencies were set by physically changing an internal crystal pack and matching antenna coil in the device, and hence could be fixed prior to actual field operations—there were no selector knobs that would accidentally turn the radio onto a wrong channel.

Thanks to all this, it was a true single-person military radio that could be operated by anyone with minimal training, and with a range of around 2–4 kilometers depending on terrain, it was perfect for front line operations.

In terms of the frequencies used, *BC-611* was built to be compatible with another portable radio, *SCR-694*, which, in its complete setup, still required a two-person crew to haul it around. This was due to the fact that it came in four bags, containing the radio unit, a set of antennas and a hand-cranked generator that removed the need of batteries. As a result, the total weight of this setup was 49 kg. When used in vehicles, like when installed on a *Jeep*, an additional *PE-237* vibrator power supply could be used to generate the various operating voltages needed by the radio.

*SCR-694* was extremely rugged, waterproof and would even float when in transport configuration. But due to its weight, it was often relegated to vehicle and semi-stationary use, for example to be used at the less-mobile field command posts, whereas *BC-611* was the workhorse for front-line use. Both of these models were essential equipment during the Allied invasion of Normandy, although due to the sheer amount of manpower and material that was thrown onto the shores of France during and after *Operation Overlord*, the previous generation, heavier and more delicate *SCR-284* was also in wide use during the invasion.

Both *BC-611* and *SCR-694* worked on the 3.8–6.5 MHz frequency range, whereas the *SCR-284* upper frequency was limited to 5.8 MHz.

At the final stages of the war, *Galvin Manufacturing* also produced the first military radio that used *Frequency Modulation*, with a type name of *SCR-300*. This backpack-sized unit offered superior audio quality and range, and was used in both Europe and the Pacific islands.

Discussion about various modulation methods can be found in TechTalk **Sparks and Waves**.

As usual, many of the great leaps in innovation that happened during wartime went straight into civilian use after the war was over. The work done on *BC-611* and *SCR-300* by *Galvin Manufacturing* is a perfect example of this, as the company changed its name to *Motorola Corporation*, and became one of the major players in the wireless business of the latter part of the 20th century.

The concept of Walkie-Talkie remains in wide civilian use, both in the 27 MHz *Citizens Band (CB)* radio, as depicted in almost any movie presenting long-haul truckers, and in the more recent personal portable transceivers, like the United States *Family Radio Service (FRS)* and Australia's *Ultra-High Frequency Citizens Band (UHF CB)*. There is no common global standard for these, however—the specifications, licensing requirements and frequencies vary across the globe, so purchasing a pair of these devices in one country and using them in another may be in violation of local laws and regulations. For example, the European *PMR* radio system uses frequencies that are in Radio Amateur use in the United States, Canada and Australia.

As all these methods use a single channel for communications, like their military ancestors from the Second World War, the connection is *half-duplex*: only one participant can talk at any one time, but an unlimited number of listeners can remain on the same channel.

*Radio control* took its first steps in military use at the beginning of the 20th century: as explained in Chapter 2: *"It's of No Use Whatsoever"*, Nikola Tesla had tried to convince the U.S. Military of the usefulness of remote control in 1898, but did not manage to gain any traction at the time. Only about a decade later, French inventor Gustav Gabet presented the first radio-controlled torpedo in 1909, and British Archibald Low was actively studying radio-controlled airplanes as early as during the First World War. Low kept on working on this subject throughout his later life, earning himself the name "Father of Radio Guidance Systems".

The value of radio control in military operations is clearly evident in our current age of satellite-controlled drones used to track down terrorists at the hotspots around the globe.

As an interesting side note, the work done in this area has spurned a very active civilian use of radio control, the history of which started in earnest in the 1950s when the transistor technology became cheap and easily accessible. Remotely flying a model airplane, or more specifically, a model helicopter, used to require reasonable skills, the lack of which would often cause very costly consequences. In contrast, the current generation of *camera drones* are automated to the point that anyone can fly them, and in the absence of a control signal, they automatically fly back to the point they took off from. It is now possible to record fascinating *High Definition (HD)* video from locations like the edges of volcanos and other dangerous places that used to be beyond the budget of all but the most dedicated producers of documentaries—another excellent example of the amalgamation of computers and wireless technology.

Radio Amateur activities, which were also briefly discussed in Chapter 2: *"It's of No Use Whatsoever"*, have at times also had significant impact on wartime operations:

At the beginning of the First World War, two British Radio Amateurs, Bayntun Hippisley and Edward Clarke, through the use of their self-built receivers, found out that the German troops were communicating with lower frequencies than the

Marconi receivers of that era could receive on, apparently expecting that nobody would be able to eavesdrop on their conversations.

Hippisley presented their observations to the Admiralty and managed to convince the military command to set up a special listening post, later known as the Hippisley Hut, on the shores of the North Sea at Hunstanton. The potential of this activity was seen as so valuable by the Admiralty that Bayntun Hippisley was given a blank check for purchasing the equipment he deemed necessary, along with the right to recruit enough personnel needed for round-the-clock operation. Most of the staff at Hippisley Hut ended up being fellow Radio Amateurs, and they built several of their listening devices on site, including novel direction-finding radios that could be used to pinpoint the source of the transmissions down to 1.5-degree precision.

Several losses to the German Naval and Airship activities can be attributed to this listening post: most notably they alerted the British Navy about the departure of the German Fleet, which led to the Battle of Jutland. Although the British lost more ships and personnel than the Germans in this particular battle, the ultimate result was that the German Navy did not venture out again in force towards the North Sea during the rest of the war, leaving it under the full control of the British Navy.

More recently, when Argentina invaded the Falkland Islands in 1982, a Radio Amateur, Les Hamilton, kept on covertly reporting on the conditions and Argentine troop locations on the island while all other communications links had been cut off. This information was crucial when the British troops started their offensive to retake the islands.

Radio Amateurs have also proven their value during peace time: when hurricane Maria devastated the island of Puerto Rico in 2017, knocking off almost all of the communications infrastructure, the *International Red Cross* asked volunteer Radio Amateurs to join their disaster recovery team in order to help in managing the information flow.

Although radio had proven to be an indispensable communications tool at war, there was one obvious drawback in relation to military operations: any transmission was detectable for anyone who wanted to eavesdrop, so in order to keep your transmissions secret, you had to encrypt the messages so they could not be understood if the enemy intercepted them.

Using encrypted messages was not a new thing—it had been used to maintain secrecy of correspondence for centuries, but what was new was the immediateness that radio provided.

Shared code books had been used to encrypt messages during the First World War, with the obvious problem of total loss of secrecy if one of the code book copies ended up in the hands of the enemy. The success of the Hippisley Hut operation was partly due to the fact that the British had managed to get a copy of the German code book, and hence were able not only to locate, but also to decipher the messages.

To avoid similar occurrences, Germany worked on mechanizing the encryption process, and at the end of the First World War, developed a system called *Enigma* for this purpose:

On a device that looked like a mechanical typewriter, you selected a set of *rotor positions* that both the sender and the receiver knew, and typed your message by using the inbuilt keyboard. The device then changed each typed letter to a new one, depending on the rotor settings, after which the resulting, gibberish-like text was transmitted by conventional means.

On the receiving side, thanks to the same rotor settings, the process was reversed and the message was turned back into its original, legible form.

*Enigma* was used in conjunction with Morse code, so it was most suitable for less time-critical communications, and was used for distributing top level instructions from the headquarters in Berlin.

Just prior to the Second World War, *Enigma* was upgraded to have five rotors for encrypting the text, which provided a total of 158,962,555,217,826,360,000 different rotor combinations. The Germans concluded that a system so complex would be impossible to break, even though it was certain that the Allied forces would eventually capture some *Enigma* devices. Having the physical device in the hands of the enemy did not matter, as there simply were too many rotor positions to choose from to randomly find the right one, and the positions in use were changed every day at midnight.

As a result, the German forces enjoyed almost complete secrecy in their internal communications until around 1940, when the first attempts to break the code started to succeed. The British understood the massive consequences of a successful code break, and did not skimp on resources: it took the effort of around 10,000 persons to finally break the *Enigma* code. The core team was working in a top-secret location at Bletchley Park in England, under the lead of a famous mathematician Alan Turing, and they were expanding massively on the prior *Enigma* decrypting work done by cryptographers in Poland.

In the end, *Enigma's* downfall came from the fact that the Germans, lulled by the apparent safety of *Enigma*, used some unnecessary, repetitive structures in their daily messages, or changed their settings only ever so slightly from the previous day, thus reducing the potential combinations by a huge factor. The British still had to use brute force to break the daily code, but the number of possible combinations changed from impossible to barely manageable.

Therefore, the technology itself did not fail *Enigma*, it was taken down by predictable human behavior.

As some of these repetitive patterns were detected and could be used as a starting point for the code-breaking effort, the British designed a machine, *The Bombe*, to methodically try to find the missing combinations, over and over again each day. *The Bombe* can be seen as the predecessor of modern computers, and due to this, Alan Turing is considered to be the "Father of Modern Computing". Even today, the *Turing Test* describes a test in which a human is having a written conversation with a computer and tries to figure out whether she's talking to a real person or a machine.

To reduce the chance of the Germans detecting that their code had been broken, forcing them to replace it by something new and potentially more complicated, statistical methods were applied on the countermeasures that were initiated based on the message intercepts: only the most decisive information in terms of the overall war situation was to be acted upon.

One can only imagine the moral dilemma facing these planners, when they had to choose on a daily basis which of the deciphered German attack plans they would foil and which would be allowed to go through. Those cases that were left untouched would still end up causing tens or hundreds of deaths on the Allied side, and sometimes close relatives of the planners were amongst the potential casualties.

As an interesting twist in this war of encrypting real-time communications, the U.S. Armed Forces came out with another, ingenious way to encrypt battlefield audio transmissions:

The Native American Navajo language was not known outside the Navajo community, so the only way to learn this language was to live amongst the Navajo people, and in the 1940s, it was estimated that only thirty persons outside the Navajo tribe could speak their language—there were no textbooks or grammars available for learning it.

Therefore, the Army recruited Navajo speakers to act as radio operators for the troops, using the Navajo language for communications.

To add another layer of complexity to this, a set of covert Navajo words were chosen for items like "tank" and "plane", including the term "crazy white man" to describe Adolf Hitler.

According to U.S. Army studies, encoding, transmitting and decoding of a three-line message could be done in 20 seconds by using Navajo "Code Talkers" as the radio operators talking to each others, compared to up to 30 minutes that would be needed if a coding machine like *Enigma* was used.

The idea of using native talkers of an obscure language for encryption purposes had already been tested on a small scale during the First World War, but it was most widely used in the battles on the Pacific islands during the Second World War. This was due to the fact that the Nazi Germany leadership was aware of this method and had sent a covert group of German anthropologists to the United States to study indigenous languages. The American leadership of the war effort did not know if these activities had borne fruit, but as they were aware of this counterintelligence activity, the decision was made to use the Navajo talkers mainly against the Japanese.

Although a system like this, which is based on a concept known as "security by obscurity", is usually deemed to be a bad approach for encryption in the long run, in this particular case, it worked perfectly. The Japanese were puzzled by what they heard and never managed to decipher Navajo language transmissions, so it remained an effective way of encryption all the way to the end of the war.

In contrast, the Japanese were able to break every other encryption method the United States used during the war.

The Allied forces, on their side, had also managed to break the Japanese codes, and like with *The Bombe*, the information gleaned from the deciphered transmissions was behind some spectacularly successful countermeasures:

By intercepting a message that indicated the precise timing of Admiral Isoroku Yamamoto's visit to the airfield of Balalae island on April 18th, 1943, the U.S. Air

Force was able to dispatch a squadron of *P-38 Lightning* fighters to intercept and eventually shoot down the *Mitsubishi G4M* bomber that was carrying Yamamoto.

This was a major blow to the Japanese forces: Yamamoto had been behind the attack on Pearl Harbor, and in practice, he was the head of the Japanese Armed Forces, second only to the Emperor in command. As a result of this successful intercept, Japan lost the key strategist behind their war effort, and never managed to replace Yamamoto with an officer of similar qualities. Also, losing a figure of such prestige right after the loss of the Guadalcanal Islands in the Pacific was a major blow to the morale of the Japanese troops: the sole reason for Yamamoto's visit was to improve the morale amongst the Japanese Forces after the recent advances by the Americans in the Pacific. Thanks to the U.S. intelligence code-breaking capabilities, the outcome was exactly the opposite.

The fascinating history of the Second World War code-breaking and encrypting efforts has made its way into several movies: an entertaining introduction to the use of the Navajo language can be found in the movie *Windtalkers*, whereas the work done by Turing and his team at Bletchley Park is depicted in the movie *The Imitation Game*.

Another, much cruder approach also exists to disrupt enemy communications: simply transmit on the same frequency as the enemy with a strong signal, and your adversary is not able to communicate. If the enemy changes to another frequency, the new transmissions can be detected in seconds via automatic scanning and blocked again.

When digital computer technology progressed to the point of being small enough to be included in transceivers, a new, indirect encryption approach, *frequency-hopping*, was taken into use by the military:

In this method, both the transmitting and receiving radios have internal timers that have been synchronized, and they share a sequence table for frequencies. The transmitter continuously jumps through the frequencies in rapid succession, and the synchronized receiver matches these frequencies by following the same frequency table.

Unless the enemy somehow has information of the sequence and timing in use, it is impossible to jam a transmission like this with a conventional transmitter. Similarly, without knowing the exact sequence, it is also impossible to listen in on a conversation, as for anyone who listens on a fixed frequency only, a frequency-hopping transmission sounds like random interference on that channel.

This approach, when properly used, provides a bullet-proof, hardware based pseudo-encrypted channel for communications.

Frequency-hopping is not only beneficial in wartime communications: it also works extremely well against any interference that appears on a fixed frequency. As this "tainted" frequency is hit only for a very short time per each sequence, its overall effect on the quality of the channel is minimized. Therefore, as we see in Chapter 9: *The American Way*, frequency-hopping eventually also found its way to everyday civilian use.

Thanks to the miniaturization of technology, dedicated military radio operators are becoming less and less common because it is now possible to provide every fighter with personal digital communications.

And instead of traditional point-to-point transmissions, all this available radio hardware on the battlefield can be used to create a *mesh network*, in which every individual radio is aware of other radios within its range and can utilize them to relay information across the mesh.

For more discussion on mesh networks, see TechTalk *Making a Mesh*.

Early attempts to create a battlefield-proof mesh networks were initiated in the United States *Department of Defense's (DoD) Joint Tactical Radio System (JTRS)* project in 1997. This has been a very ambitious program, which has suffered many cost and schedule overruns, as the expected results and the level of available technology have not converged yet. The overall cost of the project is counted in billions of dollars, with one cost estimate for an individual radio being 37,000 dollars. Despite all the problems, the program is still ongoing, and on a techno-logical level, it appears to aim at joining *mesh networks* with another emerging technology, *Software Defined Radio (SDR)*. Recent rapid development in SDR technology offers hope for JTRS to actually meet its goals, eventually unifying the radio architecture of the *United States Armed Forces (USAF)*.

Software Defined Radio is explained in TechTalk *The Holy Grail*.

Last but not least, disseminating cleverly constructed and detailed propaganda to the masses, both military and civilian, became an important aspect of the com-munications warfare during the Second World War.

The Axis forces made daily transmissions that were meant to reduce moral amongst the Allied forces: Germany's Lord Haw-Haw, an Irish collaborator who spoke with a peculiar English accent, was on the air for practically the whole duration of the Second World War, bombarding the Allied troops with propaganda, while in the Pacific theater, Japan's Tokyo Rose, a team of English-speaking Japanese women, "entertained" American forces, interlacing popular American music with dire threats of impending doom and gloom.

Although the true purpose of such transmissions was obvious, there was enough real information sprinkled in, keeping the listeners interested—thanks to the severely restricted news flow from domestic sources, especially regarding military operations that had not been successful, both soldiers and civilians were keen to find out any additional snippets of information regarding their comrades or loved ones, and listening to these enemy transmissions provided an alternative to the highly censored messages from their own government.

Creating the right mix of pure propaganda and credible information was an artform, and it turned out to be effective enough to force the Allied Command to improve their own information flow, even when the news was negative.

In the end, contributing to these propaganda attempts did not bode well for the participants:

The main person behind Lord Haw-Haw, William Joyce, an Irish fascist who had fled to Germany in 1939, was hanged for treason after the war, and one of the alleged women behind Tokyo Rose, an American-Japanese Iva Toguri D'Aquino, ended up being jailed for treason when she tried to return to the United States. She was eventually released after ten years in prison when an appeal concluded that there was not enough evidence to prove her participation in the Tokyo Rose transmissions, but the shadow of this potential connection hung over her all the way to her death at the age of 90.

Entities like Lord Haw-Haw and Tokyo Rose were made possible by the fact that a new, unidirectional form of wireless communications, *broadcasting*, had become prevalent during the years between the two world wars.

This *Golden Age of Wireless* had a profound effect in our society, and the fascinating story that led to this first consumer electronics boom is the subject of the following chapter.

# Chapter 4
# The Golden Age of Wireless

The son of a Canadian Reverend, Reginald Fessenden was only fourteen years old when he was granted a mastership in mathematics at the Bishop University in Quebec, Canada, allowing Fessenden to teach boys in his own age group at the accompanying Bishop's College.

Despite his abilities and apparent completion of the necessary level of studies for graduating, he left the university at the age of eighteen, with no formal graduation, and moved to work as a teacher on the island of Bermuda.

His personal focus, however, was squarely on the ongoing electricity revolution that was in full swing in the mainland United States. And after just two years, his curiosity finally got the better of him: he moved to New York with a clear goal of working for the "genius himself", Thomas Edison.

Even though Edison at first did not approve of Fessenden's unconventional academic background, Fessenden's insistence on "whatever job will do" made Edison give him a chance, and he got a semi-skilled tester position in one of Edison's companies. During the following four years, his dedication paid off and he swiftly rose within the company, becoming the Chief Chemist, until in 1890, Edison's financial problems forced him out, along with many other prominent engineers.

This minor setback did not stop Fessenden: just like Tesla, he was briefly employed by George Westinghouse, and with his newly acquired practical background in electrical engineering, he became the head of the electrical engineering department at the University of Pennsylvania.

He worked on a practical demonstrator for the newly discovered radio waves, but felt that he had lost the race when Marconi eventually announced a successful demonstration in 1896.

Fessenden managed to build a working setup based on the conventional spark-gap approach, and one day while he was experimenting, the Morse key on a transmitter got jammed, causing his receiver to produce a continuous, howling sound. After investigating this further, Fessenden theorized that radio transmissions could somehow be modulated to also carry voice, and researching this possibility became his main focus for years to come.

© Springer International Publishing AG, part of Springer Nature 2018
P. Launiainen, *A Brief History of Everything Wireless*,
https://doi.org/10.1007/978-3-319-78910-1_4

When he was passed over for a promotion at the university, he left academia and started looking for private funding that would enable him to continue his research in this area.

Fessenden's first financial breakthrough came in 1900, when he managed to get a contract with the *United States Weather Bureau* to change the Bureau's weather reporting setup from telegraph lines to radio. The wording in his contract allowed him to retain ownership of anything he invented, and he made several improvements to signal detector technology during this time.

But his personal goal continued to be the transmission of audio over radio waves, and at the end of 1900, he finally succeeded.

It's worth remembering that his early audio transmission equipment was still based on the rudimentary electro-mechanical spark-gap technology, as the invention of solid-state electronic components had not yet happened. Therefore, the transmission of voice instead of Morse code was a major accomplishment, considering the extremely primitive technology he had to work with.

Over the years, Fessenden turned out to be a very prolific inventor, having more than 500 patents to his name. He was the first to achieve two-way Morse communications across the Atlantic Ocean in 1906, surpassing the one-way connection made by Marconi five years earlier. This work introduced him to the effects of the *ionosphere*, the layer in the upper atmosphere that was reflective to the main frequencies in use at the time. He noticed that the reach of his transmissions varied greatly over the period of a year and also daily, due to the effects of *solar radiation*. He even managed to perform an unintended transatlantic audio transmission: a test that was targeted to a receiver in the United States was by chance heard in Scotland. Fessenden wanted to repeat this in a controlled setup later in the year during optimal ionospheric conditions, but unfortunately a winter storm destroyed the antenna of the Scottish receiving station.

The success in creating audio transmissions was not Fessenden's only groundbreaking invention: his theory of the *Heterodyne Principle* in 1901 was ahead of its time, three years before there even were electronic components that could prove the theory in practice. This principle, in its applied form of *super-heterodyne* receiver construction, is still the main radio receiver technology in use, and it is described in TechTalk *Sparks and Waves*.

Fessenden's early advances in audio transmissions drew interest from financiers, and under the newly created *National Electric Signaling Company (NESCO)*, he kept working on a continuous wave transmitter technology in order to replace the existing spark-gap systems. During this process, he transferred some of his existing, essential patents to *NESCO*, filing also several new patents in this area on behalf of *NESCO*, thus building a significant portfolio of intellectual property for the company.

While he was pushing this new approach, he was belittled by some of the other wireless pioneers of the era, who strongly believed that spark-gap systems were the only possible way to create radio waves with sufficient intensity.

And in practice, creating high-frequency generators through electro-mechanical equipment proved to be very complicated: the successful demonstrators were able to generate only very low transmission power, which in turn limited the distance covered by Fessenden's early continuous wave transmissions to roughly 10 kilometers. Fessenden had subcontracted this work to the *General Electric Corporation (GE)*, a company that had its roots in the merging of the multitude of companies set up by Thomas Edison. Fessenden's contract at *GE* was performed under the supervision of Ernst F.W. Alexanderson, who continued this line of work when the Fessenden's contract expired, eventually getting his name in the history books after creating a series of very successful *Alexanderson Alternators*, electromechanical continuous wave transmitters that were powerful enough to reach over the Atlantic.

But Fessenden was not too worried about the limited range in the early years of the 20th century—he understood the vast potential in being able to use audio for the modulation, and dealing with power was important only after the modulation issue was satisfactorily solved.

Fessenden achieved good progress in his numerous demonstrations, including the first transmission of music around Christmas of 1906, during which he also sang and played the violin himself. He also used a phonographic record as an audio source for his transmission.

Thanks to this successful demonstration, Fessenden can be seen as the first implementer of the *broadcasting* concept—the unidirectional transmission of programming to listeners.

And by mixing his commentary with pre-recorded music, Fessenden also unknowingly became the world's first radio DJ.

A couple of days earlier Fessenden also demonstrated the interconnection of the wired telephony network and his wireless audio transmission system at the Brant Rock, Massachusetts facility, thereby joining these two communications technologies.

Unfortunately, as often is the case with truly pioneering companies, *NESCO* failed to provide the kind of return on investment that its financiers had expected, and this caused considerable friction between Fessenden and other shareholders. Their early attempt to gain foothold in the U.S Navy's radio business, as described in Chapter 3: **Radio at War**, was unsuccessful, and later in the history, the overall situation was not helped by Fessenden's wide business activities outside of the *NESCO* company structure. Hence, a long, complex legal battle between him and his financial backers ensued, eventually putting *NESCO* into receivership.

But shortly thereafter Fessenden's luck finally turned: the newly invented vacuum tube technology became available, providing a viable alternative for the generation of high-power continuous waves. As a result, the patents that Fessenden had filed both in audio modulation and receiver technologies finally started to bear fruit: *NESCO's* patent portfolio was sold first to *Westinghouse* and later ended up as intellectual property of the *Radio Corporation of America (RCA)*.

Fessenden disputed part of this deal, and in 1928, only four years before his death, *RCA* settled the case with a considerable cash payment. For *RCA*, this ended up being a very good deal, as the company took full advantage of the technology it gained and went on to become one of the largest corporations of the mid-20th century.

Despite the numerous setbacks and the initial bad luck in his business endeavors, Fessenden's gradual advancements in continuous wave technology over a period of more than twenty years were notable in the history of radio transmission technology: they were the crucial steps that led to the broadcasting revolution, which was spearheaded by *RCA*.

The story of *RCA* starts right after the First World War was over:

At the beginning of the war, one dominant company managed the wireless transmissions across the Atlantic Ocean—the *Marconi Wireless Telegraph Company of America*, which was part of the Marconi's business empire in Britain.

The assets of this company were taken over by the U.S. military during the war, as was required by the law at the time, but instead of returning them back to Marconi when the war was over, the company's assets were forcibly bought out in a series of moves that can only be described as highly opportunistic and protectionist interchanges, aimed at squeezing a foreign company out of the American market.

The reason for this sell-out was a unique technical detail that made Marconi's position very vulnerable:

The major owner of *RCA* was the *General Electric Corporation (GE)*, which, as discussed above, also manufactured Alexanderson Alternators. These were the essential part of the long-range transmitters manufactured by Marconi and sold worldwide, and there were no other means of producing high-power continuous waves at the time. Therefore, when *GE* suggested that they would only keep on selling these alternators to Marconi for use outside of the United States in exchange of his American assets, Marconi was effectively left with no other option than accepting the takeover of his American subsidiary.

As a result, the assets of the *Marconi Wireless Telegraph Company of America* became the initial core technology of *RCA*. The ultimate irony in this case was the fact that the rapidly advancing vacuum tube technology made Alexanderson Alternators obsolete in less than five years.

The U.S. Army and Navy worked as the midwives of the newly established company, gaining board seats, and most importantly, through the forced pooling of patents during the war, helped to defuse many potential patent disputes around the wireless and electronic component technology between *GE* and other major companies. Thanks to the resulting strong patent portfolio and immensely wealthy backers, *RCA* was in an optimal position to gain from the post-war socio-economic boom, making the story of *RCA* one of those "in the right place at the right time" successes:

Radio technology had matured, high-power audio transmissions were now possible, and the cost of manufacturing radio receivers had fallen to a level that made them suddenly affordable to the masses. As a result, radio receivers became a "must have" apparatus, and households went out and bought them by the pallet-load.

The demand hit the *hockey stick curve* phase that has been seen so many times thereafter when new, highly useful inventions become affordable to all: suddenly sales will skyrocket, and those companies who have the best products available will be able to sell as many as they can manufacture.

*RCA* was in the middle of this consumer storm, and with one of the great business leaders of the time at its helm, David Sarnoff, the company rapidly became the behemoth of all things wireless, creating the first globally recognized communications empire.

David Sarnoff was a Russian immigrant who had earlier been employed by Marconi as a wireless operator of Marconi's Nantucket radio station. He was on duty when the fateful message was received:

S.S. Titanic ran into iceberg, sinking fast.

Sarnoff claimed to have been relaying messages related to the rescue mission uninterrupted for the next 72 hours.

During the four years after the RMS *Titanic* accident, Sarnoff rose quickly among the ranks of Marconi's American business, and according to a story that has been widely referred to as the *Case of the Radio Music Box Memo*, he tried to sell his idea of the *broadcasting model* to the management of the *Marconi Wireless Telegraph Company of America* in 1916, stating that:

I have in mind a plan of development which would make radio a 'household utility' in the same sense as the piano or phonograph. The idea is to bring music into the house by wireless.

He got no traction for his proposal, and when the First World War broke out, it put a complete freeze on civilian radio development for the entire duration of the war.

When Marconi's American assets were transferred over to *RCA* after the war, most of the staff came as part of the deal, including David Sarnoff. In this new context, Sarnoff finally had the support he needed to continue with his bold pre-war plans:

Sarnoff understood that by disseminating entertainment, news and music to the masses, there would be a huge demand for the devices *RCA* was producing, and he successfully persuaded the budding *RCA* to provide the funds needed to turn his concept into a large-scale, multi-faceted operation that covered not only the manufacturing of radio receivers, but also the creation of the material to be transmitted. His actions started the revolution that turned the established two-way communications system into the multi-billion, one-way broadcast imperia that we have today.

In 1921, he organized the live broadcasting of a heavyweight boxing match, and the number of listeners that night was calculated in hundreds of thousands. This was a far bigger audience than could ever be achieved in a physical venue.

The market of easily reachable, huge audiences was born.

Events like this showed the potential of this new approach, sales soared, and *RCA*'s march to market domination was secured: *RCA* became the American catalyst of the *Golden Age of Wireless*, and the airwaves were filled with all kinds of entertainment, from music to live sports broadcasts to up-to-date news bulletins. It also created as a totally new experience: *radio theater* with soundscapes that tickled the listeners' imagination.

Sometimes these *radio plays* succeeded even beyond the producers' wildest dreams:

On the calm Sunday evening of October 30th, 1938, casual listeners to the *Columbia Broadcasting System* network were entertained with smooth live music "from Park Plaza Hotel in New York".

Soon, however, the music started to be intertwined with news bulletins: first a normal weather report, then a report about strange telescopic observations of bright flashes on the surface of the planet Mars.

An interview with a perplexed astronomer was heard, after which the music continued.

Then another news bulletin reported on sightings of multiple meteorites that had been observed in New Jersey.

More music and ordinary news flashes followed.

But soon the contents of these seemingly live news bulletins took an alarming turn: strange cylindrical objects were described on the ground in Grover's Mill, New Jersey, surrounded by an ominous sound. These reports were followed by eyewitness descriptions of "strange creatures" crawling out of the cylinders.

The reporting became more and more ad hoc and panicky, sounds of sirens and shouting were heard on the background, and finally the live transmission was abruptly cut off.

When the transmission continued, "a technical problem with the field equipment" was blamed for the interruption, but after another brief interlude with music, there was a grave report of multiple deaths caused by "the invaders from Mars".

Declaration of Martial Law in New Jersey was announced.

The army was reported to attack these invaders, followed immediately by live eyewitness stories of huge casualties caused by the Martians, while at the same time, more and more cylinders were reported to have landed.

A highly emotional speech by the "Secretary of Interior from Washington" then followed, describing a "horrendous attack" against the people of Earth.

Bulletin by bulletin, the situation got worse: bombers and artillery were reported to counterattack the Martians, with poor results, and eventually reports were flooding in about how the neighboring city of New York was also being taken over by the Martian invaders.

But if anyone listening to this in New York or New Jersey had bothered to look out of their window, there were no ray-gun touting Martians anywhere to be seen.

Despite this potential for easy verification, some listeners panicked as the fake bulletins kept on piling up with ever-worsening reports.

At this point, the first announcement about the program being a theatrical pre-Halloween prank was made, but apparently not everyone got the message.

The story continued with the same gravity, until finally, after reports of widespread destruction of New York, the Martian invaders suddenly died: apparently, they had no immunity against our common, earthly germs.

The battle was over: Earth 1, Mars 0, with just a little help from viruses.

In real life, though, any potential Martians were still solidly on their home planet, and the program was an ingenious adaptation of the novel *The War of the Worlds* by H. G. Wells—a fact that most certainly some of the listeners had already figured out, as the storyline and the descriptions of the Martians were eerily identical to the ones in the book they had read.

But apparently many others did not realize this, and the "reality show" approach of the show had scared hordes of people who had tuned into the *CBS* channel. Quite a few of the listeners, having joined in the middle of the program during its most intense moments, had missed the first announcement of it being just a simple prank.

As a result, listeners around the United States had taken what they heard as a depiction of a real event, and the director of the show, a young, ambitious man called Orson Welles, suddenly found himself in very hot water: even some suicides were blamed on the show.

In a hastily set up press conference the following morning, Welles somehow managed to talk himself and his company out of either being sued or losing its broadcasting license. Expressing deep remorse, he started his comments by saying:

We are deeply shocked and deeply regretful of the results of last night's broadcast.

Welles then referred to the "unforeseen effects" of the show, and emphasized that several announcements about the show being just a radio theater play had been made during the broadcast.

As a result of the nation-wide media circus around the effectiveness of show, Welles was propelled on to a very successful career in Hollywood, which included the creation of one of the greatest cinematic masterpieces, *Citizen Kane*.

*The War of the Worlds* concretely showed the widespread power of broadcasting, both good and bad. Radio was no longer useful only as a communications tool for two-way information exchange, it was also a unidirectional, powerful delivery media, able to reach large swaths of the population in real-time, and it was a tool that could be used either to sooth or to inflame the emotions of the masses.

And this immense power could also be used for purely nefarious propaganda purposes, as was discussed in Chapter 3: **Radio at War**.

All this was made possible through the rapid development of solid-state electronic components, which enabled the receiver technology to both go down in size and become cheap enough for the common man to purchase.

The cheapest radio receiver, the *crystal receiver*, had been around since 1904. It was suitable both for spark-gap and continuous wave transmissions, but needed the listener to use headphones in order to hear anything. It was also very insensitive, forcing the listener to be relatively close to the transmitter in order to receive the signal.

In the 1920s, vacuum tube based, affordable units become available, thanks to the rapid developments in miniaturization achieved during the First World War years. As these units now had inbuilt amplifiers and loudspeakers, listeners no longer needed headphones. Due to the active electronics, together with the application of *super-heterodyne* technology, the range of reception was hugely increased, thus multiplying the number of stations that were available at any given location.

TechTalk *Sparks and Waves* describes the benefits of solid-state electronics and the super-heterodyne technology in detail.

For the first time, the whole family could now gather around the receiver for a collective listening experience: a new family pastime was born.

The increased demand led to mass production of receivers, which, together with strong competition between manufacturers, quickly brought prices down. The world was witnessing the first consumer-electronics boom—a phenomenon which has since been repeated umpteen times as new technologies have matured for mass adaptation.

The added complexity of receivers, together with the limited lifespan of vacuum tubes, meant that the receivers required maintenance from time to time. This, in turn, created a totally new service industry that was required to keep the ever–expanding pool of consumer devices in working order.

On the broadcasting side, the rapidly growing mass audience expected an uninterrupted stream of entertaining content, and in 1920, *Westinghouse* became the first, government-mandated broadcaster in the United States. *Westinghouse* had also managed to become one of the majority owners of *RCA* due to the fact that they had bought the patents for the *super-heterodyne* technology, and *RCA* simply had to have access to it in order to be able to produce competitive radio receivers.

Originally, broadcasting content generation was seen simply as a necessary cost item needed to support radio equipment sales, but as the number of broadcast stations mushroomed and the costs to maintain them exploded, radio advertising became a lucrative source of funding. A monopoly for advertising-based radio was initially granted for *American Telephone and Telegraph (AT&T)*, but such a vast and continuously growing market was soon deemed inadvisable to be handled only by a single company. As a result, the roots for huge networks like *Columbia Broadcasting System (CBS)* and *National Broadcasting Company (NBC)*, the broadcasting arm of *RCA*, were laid out. These networks rapidly expanded across the United States, soon covering all major urban areas.

As national networks were built up and formerly individual stations were networked and syndicated together, the audience potential for this new medium grew beyond anything that had been experienced before. For the first time, it was possible to follow news, events, religious sermons, sports and other entertainment on a national scale, in real-time, as the events unraveled. Radio became a unifying medium, providing an endless stream of information and entertainment, either as a background or as a centerpiece of concentration.

Radio was also the first socio-economic equalizer, as anyone, however far from the metropolises of the era they lived, suddenly had access to the same stream of information as educated city-dwellers. Some newspapers panicked and even chose not to publish listings of radio programming, fearing that they would lose their business model: how could the printing press possible compete with something that was instantly accessible and up-to-date with the latest events?

The potential for national coverage meant enormous audiences. Already in 1924, there were over three million receivers in the United States alone—about one in ten households had bought one, and cafés, restaurants and other public places saw the radio receiver as a magnet for customers.

At the same time, the number of radio stations had grown already to around 500.

Considering that Fessenden's first audio transmissions had happened only 18 years earlier, and the first industrially produced vacuum tubes were manufactured in France only nine years earlier, this presented a growth of truly astounding proportions.

On the technology side, *RCA*, which had also become one of the major manufacturers of vacuum tubes, saw its component production grow from 1.25 to 11.35 million units between 1922 and 1924.

When the *Great Depression* hit the United States in 1929, most businesses suffered, except for the broadcasting industry: radio was seen as a source of cheap entertainment, with no need to buy tickets or to pay for travel, so it was no wonder that four million additional radio receivers were sold between 1930 and 1933.

Having a nationwide network with advertising rights was equal to the right to print money: continent-wide programming was made available, complemented by hundreds of smaller, independent and locally operating radio stations.

Advertising was an important catalyst for the rapid expansion of the new medium, as it provided a simple and sustainable funding model. To reach a guaranteed audience of millions of listeners to the most successful shows and sports events, the advertisers were keen to compete over the available slots, and the nationwide networks were happy to sell airtime to the highest bidder. In the United States, the joining of radio and advertising was a marriage made in profit heaven.

On the other side of the Atlantic, a very different approach to the early deployment of this new technology was taken:

The first national broadcaster in the United Kingdom was set up through the creation of the *British Broadcasting Company (BBC)* in 1922, and one of the founders of the *BBC* was none other than Guglielmo Marconi himself. In the following year, an annual radio license fee of ten shillings was introduced under the *Wireless Telegraphy Act*.

To put this in perspective, the average weekly pay for a manual worker was two pounds and twelve shillings, so the cost of enjoying radio in Britain was by no means excessive in relation to the existing income level, yet it allowed the British approach to function without direct advertising.

The fact that the funding came through an indirect tax levied by the government put the *BBC* in a much more independent position compared with the advertising-based radio networks, and perhaps due to this fundamental difference, the *BBC*, which was renamed the *British Broadcasting Corporation* in 1927, is still, almost a hundred years after its creation, seen as one of the most balanced and impartial news sources in the world.

Great Britain has maintained its license structure, although today it is a license for television, not radio, and it costs about 200 dollars per year in 2017, covering up to 15 receivers per license in the same household or company context.

Therefore, the average British household pays about 50 cents per day for the continued independence of the *BBC*.

The combination of an existing mass-market and the ongoing rapid developments in technology created a virtuous renewal cycle for the manufacturers. High quality audio transmissions became possible due to the move to *Frequency*

*Modulation (FM)* at the end of the 1930s, giving the customers an incentive to upgrade their receivers accordingly: the now ubiquitous *FM radio* was born.

Frequency Modulation is explained further in TechTalk *Sparks and Waves*, and the relationship between higher frequencies and the corresponding potential for better audio quality is discussed in detail in TechTalk *There is No Free Lunch*.

Thanks to the advances in miniaturization, the first portable radios became available in the 1930s. Smaller sizes also made it possible to fit receivers in cars, enabling access to entertainment during the long, dull drives, that were, and still are, very common in the continental United States.

Music laced with direct advertisements were the main ingredients of programming for many stations, but advertisers also used more subtle approaches.

Sponsored, long-running radio plays that referred to the sponsor either directly or indirectly were very successful, providing a powerful hookup between advertised products and customers. These serialized shows were played daily, leaving the listeners with a cliffhanger, eagerly anticipating the next part, which, naturally, was laced with more references to the sponsor.

Most of the soundscapes created for these shows were produced live during the actual broadcast, and were amazingly complex and innovative. They conveyed the sonic imagery supporting the storyline, nursing the listeners' imagination, and hence everyone had their own, personal experience of the show.

But with the introduction of television, the enchantment of moving images totally reshaped the experience of the audience—everyone saw and shared the same imagery: it was like in a cinema, except that it happened right there in your living room.

Radio theater is a rare exception to the programming these days, as the mesmerizing effect of the moving image on television offers a much more easily digestible, although also much more passive experience.

Hence, the broadcast model continued through this new, visual medium, but the focus of new developments, along with big advertisement budgets, rapidly moved over to television. The progress that followed, starting from the 1950s, has been aptly summarized in a song by the band *The Buggles*: "Video killed the Radio Star".

The emergence of television created yet another, totally new market for manufacturers and companies providing maintenance for the devices, while the additional power of moving images lifted advertising to a totally new level of persuasion.

As a result, radio programming budgets crumbled and the value of nationwide syndication faded. The structure and the expectations of the audience of radio programming changed radically: radio's focus turned increasingly to news, talk shows, disseminating local information and especially music, catering for those moments when your main focus had to be somewhere else, like during your daily commute.

In the end, it was bands like *The Buggles* that helped to save the radio through the new emphasis on music content, driven by the technological enhancement that further improved the listening experience: the addition of stereophonic sound to FM radio.

The change to stereophonic transmissions was done in an ingenious way that did not invalidate the millions of existing monophonic receivers that were already out on the market. This was achieved by cleverly utilizing the wide channel bandwidth of FM radio: the analog sum of the left and right channels of a stereophonic signal is sent exactly like before, resulting in the basic, fully backwards compatible monophonic sound. But in addition to this, the analog difference signal of left and right audio channels is used to modulate a 38 kHz *carrier signal* that is then added on top of the normal monophonic transmission.

As the channel width for FM is 100 kHz and the monophonic audio is limited to 15 kHz maximum frequency only, there is ample room to inject such an additional 38 kHz carrier signal without any changes to the underlying system.

Monophonic receivers won't react to this extra information—as it is embedded into a 38 kHz carrier signal, normal loudspeakers and headphones can't even reproduce sounds with such a high-frequency, and even if they did, it would be way beyond the range of human hearing. But new stereo receivers extract and *demodulate* this additional carrier signal, recreating the original left and right signals through simple analog addition and subtraction between the two parallel signal sources, L+R and L−R.

Further discussion on the concept of bandwidth can be found in TechTalk *There is No Free Lunch*.

Apart from the *Radio Data System (RDS)* extension, offering automatic switching to channels sending traffic announcements and the embedded transmission of station identifier information, FM radio has remained essentially the same since the early 1960s. But with the advent of digital technology, audio broadcasting is now facing another major technological change due to the introduction of *Digital Audio Broadcast (DAB)* transmissions.

Concepts like *analog* and *digital* are discussed in TechTalk *Size Matters*.

The biggest obstacle for the deployment of this new digital extension is the fact that DAB is not compatible with the billions of FM receivers that are already out there, so the progress has been very slow so far. The cheapest DAB receivers have come down to 30-dollar level in Europe, but a traditional analog FM radio can still be bought for less than ten dollars, and FM receiver functionality is often even embedded in your cellular phone.

What is even worse, DAB went through an improvement step to DAB+, which was not compatible with the first-generation DAB receivers, making all early adopters very unhappy. Currently over thirty countries have either experimented with DAB or have regular transmissions based on this technology, but traditional analog FM radio has shown great resilience so far, thanks to the universal, established and cheap end-user technology.

A good case study for DAB is the United Kingdom, which, together with Sweden and Norway, was one of the three earliest experimenters with DAB transmissions in 1995. The listener base for DAB transmissions has remained relatively static, even though the London area already had fifty DAB stations by 2001.

Despite all these potential issues, Norway, a country that was the first to introduce DAB broadcasts, switched off analog FM radio and went all DAB in December 2017.

This first attempt of FM radio switch-off in Norway will be keenly followed by all other broadcasters around the world. Norway is a rich country with good infrastructure and high average standard of living and hence should be the easiest place for such a fundamental transition.

As a reference, there are no plans to move to DAB in the United States, where *Sirius XM* satellite radio offers multi-channel digital distribution across the whole country, and an alternative, hybrid analog/digital version, *HD radio*, is being introduced.

In HD radio, the extra bandwidth of an FM radio channel is used to embed a simultaneous digital signal on top of the conventional analog transmission using the same kind of backwards compatible approach as for embedding stereophonic audio. This data stream can offer the same audio in digital format, while also allowing a limited set of multiple parallel digital channels for alternative content.

Therefore, HD radio channels still function with normal FM receivers. In this sense, it is a more practical upgrade to FM radio than DAB, but whether it really catches on remains to be seen.

The technical reasons behind the move to DAB and HD radio from the broadcasters' point of view are the same as for digital television, and they are discussed in detail in Chapter 5: *Mesmerized by the Moving Image*.

Digital data stream has some issues that can make listening to it a much more annoying experience than traditional FM radio:

On analog FM radio, even when there is a low signal strength and lots of interference, it is usually possible to comprehend what is going on. Most of the time there is also continuity: the signal quality may be poor, but it does not totally drop out for seconds at a time.

With digital, you tend to either have a perfect signal or annoying "holes" in reception, sometimes accompanied with loud metallic screeches, as the received, digitally encoded signal is incomplete and is therefore decoded incorrectly. This tends to be more of a problem when you are on the move, which is unfortunate, as listening to radio is one of the major activities while driving.

For the satellite-based *Sirius XM*, these reception issues tend to be less prominent, as the signal is always coming "from above", with an angle that depends on your latitude. As your satellite antenna is usually on the roof of your car, it keeps on having a line-of-sight connection to the satellite above. Naturally, stopping under a bridge or driving into a multi-level car park may cut off the signal, but otherwise the reception quality tends to be excellent.

Personally, the only time I had issues with *Sirius XM* reception outside of the above-mentioned, literally "concrete" examples was when I was driving through a thick forest near Seattle: the trees were high enough to cut my line-of-sight to the satellite over the Equator while I was on such a Northern latitude, and thick enough to suppress the incoming microwave signal.

Apart from the differences in poor or mobile reception situations, selling the promise of DAB to the public will be much harder than with digital television: common stereo FM transmissions are usually "good enough" for most consumers, whereas with digital television, the improvement in picture quality is much easier to perceive, and the emergence of HD audio as a backwards-compatible method may prove its value in the long run.

In addition to that, thanks to the full flexibility available for defining the bandwidth used for individual digital channels, the switch to digital audio on the broadcaster side can also be done in a way that defies logic—a case in point is the DAB-pioneering United Kingdom, where some channels chose to use a very low bit rate and even a monophonic signal, both of which caused the DAB audio to sound utterly inferior compared to traditional analog FM broadcasts.

For both DAB and HD audio, the jury is still out.

But as the next revolution in broadcasting was happening with the advent of television, the stereophonic audio experience, whether analog or digital, was no competition to the power of the moving image: television rapidly took the upper hand.

# Chapter 5
# Mesmerized by the Moving Image

A combination of an inquisitive young mind, lots of spare time and inspiring reading has propelled many inventors into their successful careers. For a twelve-year-old youngster called Philo T. Farnsworth, this kind of fruitful cohesion happened after his family moved to a large farm in Rigby, Idaho, in 1918.

While investigating his new home, Farnsworth found a pile of books and magazines about technology in the attic, devouring their contents during his spare time.

The house also had a feature that was novel to Farnsworth and greatly tickled his curiosity—a rudimentary generator of electricity, supplying the electric lights of the farm. This was a huge improvement over the simple oil-lamp lighted log cabin Farnsworth had lived in in Utah before the move.

Farnsworth learned the quirks of the somewhat unreliable generator and, to the joy of his mother, proceeded to install a salvaged electric motor to run the manual washing machine of the household.

The books and magazines that Farnsworth had found contained references to the concept of television, although no functional wireless demonstrators were yet available. The state-of-the art in wireless at the time were simple audio transmissions, but the idea of transmitting pictures over the air was speculated on widely in technology-oriented journals.

The current idea for television at the time was based on a mechanical concept called the *Nipkow disk* that had been patented in 1884. In a Nipkow disk-based system, the image is sliced into concentric arcs by a circular plate that has holes evenly placed along its radius, and the light at each arc is used to modulate an electronic light sensor. Then by reversing the process on the receiving side in sync with the transmitter side, the dynamically changing view that is projected on the transmitting Nipkow disk can be reproduced.

This was a clumsy approach at best: having a large disk spinning at high speed would make the setup noisy and prone to spectacular mechanical crashes.

The very first functional Nipkow disk-based system was created by Russian Boris Rosing in 1907, and he continued actively working on his designs over the years. Rosing got several patents for his system and his demonstrator was presented

© Springer International Publishing AG, part of Springer Nature 2018
P. Launiainen, *A Brief History of Everything Wireless*,
https://doi.org/10.1007/978-3-319-78910-1_5

in an article published by the *Scientific American* magazine, complete with system diagrams.

Unfortunately, Rosing became one of the victims of Joseph Stalin's purges, and died in exile in Siberia in 1933.

Farnsworth understood the fundamental limitations of a mechanical approach, and started drafting a fully electronic version while he was still in high school. His conceptual design consisted of horizontally scanned, straight lines that would be stacked and repeated in rapid succession to form a continuous stream of individual image snapshots, *frames*. If these were shown fast enough in succession, they would give the impression of a moving image.

According to Farnsworth, he got the idea for *scan lines* while plowing the family fields.

Thanks to the fact that he discussed his idea with his high school chemistry teacher, giving him a schematic drawing of his planned system, he inadvertently created a valuable witness who would later help him win an important patent dispute against the broadcasting behemoth *RCA*.

Another inventor, Karl Braun, had already solved the solid-state display side of the television concept in 1897 with the *Braun tube*, the predecessor of all *cathode ray tubes (CRTs)*, which were fully electronic and fast enough to create an illusion of seamless movement. Therefore, no moving components would be needed on the display side of any television solution. Even the early version by Rosing had used a CRT for the display side.

Braun was also one of the early pioneers in wireless technology, having made many advancements in the area of tuning circuitry. This earned him the Nobel Prize in Physics in 1909, shared with Guglielmo Marconi, as discussed in Chapter 2: *"It's of No Use Whatsoever"*.

Removing mechanical parts from the transmitter side of a television system needed a totally new approach for image processing, and Farnsworth's solution for this was called an *image dissector*:

A traditional camera lens system is used to project an image on a light-sensitive sensor plate, causing each area of the sensor to have a tiny electrical charge that is directly proportional to the amount of light being exposed on the corresponding location on the plate. By encasing this sensor into a vacuum tube-like structure with electric grids in both horizontal and vertical directions, an electron beam can be deflected along the sensor's surface, resulting in a constantly varying electron flow that is proportional to the amount of light arriving at each scanned location.

As the image dissector is fully electronic, this process can be repeated with high enough speed to cover the full sensor area several tens of times per second, and hence the successive set of still images becomes rapid enough to be perceived as continuous movement by the human eye.

By using the output of an image dissector as a modulating signal for a transmitter and embedding suitable timing control information into the signal, the stream of captured still images can thereafter be reconstructed at the receiving end, and the existing CRT technology could be utilized for displaying the received transmission.

Farnsworth's functional image dissector prototype was the first device providing the principal functionality of a *video camera tube*, and paved the way towards high-quality, easily maneuverable television cameras.

It took Farnsworth several years to finally turn his idea into a system that actually was able to transmit an image, but his faith in the huge financial potential of his invention was so strong that he even asked for an honorable discharge from his early employer, the prestigious *United States Naval Academy*. This ensured that he would be the sole proprietor of any patents he would file in the future.

Farnsworth was able to convince two San Francisco philanthropists to fund his research with 6,000 dollars, and this injection of money, equivalent to about 75,000 dollars today, enabled him to set up a lab and concentrate fully on his television idea.

With a working prototype of an image dissector, he succeeded in sending the first, static image in 1927. When the stable image of a simple, straight line of the test transmission picture appeared on his receiver's CRT, he had no misunderstanding of what he had just proved, commenting:

*There you are – electronic television.*

Farnsworth also showed that he had a good sense of humor: as his financial backers had constantly pushed him to show financial return for their investments, the first image he showed them was an image of a dollar sign.

In 1929 he managed to remove the last mechanical parts from his original setup and transmitted, among others, an image of his wife—the first live television transmission presenting a human subject.

But the competition was heating up, and it came with deep pockets:

Despite having a working electronic image dissector system, with relevant patents to back up his claims, Farnsworth ended up in a bitter dispute with *RCA*, which tried to nullify the value of Farnsworth's patent. *RCA* had had its own, big budget development ongoing for video camera solutions that followed the same principle as Farnsworth's version, and David Sarnoff wanted to ensure that *RCA* would be the patent owner, not the patent licensee.

Sarnoff had tried to buy rights to Farnsworth's patents for the lofty sum of 100,000 dollars and an offer to become an employer of *RCA*, but Farnsworth preferred his independence as an inventor, aiming to profit from licensing instead. This was against Sarnoff's fundamental idea of avoiding licensing fees, and he started a litigation process against Farnsworth. In court, the *RCA* lawyers openly belittled the idea that a farm boy could have come up with a revolutionary idea of this magnitude, but despite the money and manpower *RCA* was able to throw in, they eventually lost the case. A crucial witness statement that supported Farnsworth's claims came from the aforementioned chemistry teacher, who could present the original image dissector schematics drawing that Farnsworth gave him several years earlier.

This important detail dated Farnsworth's work on the subject further back in time than the work done in *RCA's* laboratories by Vladimir Zworykin, who also had filed several early patents in the area of television technologies.

What made *RCA*'s case even weaker was the fact that Zworykin could not provide any functional examples of their claimed earlier works. His patents were also overly generic. Farnsworth, on his side, had a working prototype that accurately matched his patent application, although it had been filed four years later than the version that *RCA* was pushing through the courts. Sarnoff saw the weak position that *RCA* was in, but decided to drag the case as far as possible, relentlessly appealing against the decision. His plan was to keep Farnsworth occupied with the process, draining his funds and reducing the remaining active time of the patent protection.

*RCA* had basically lost the case already in 1934, but it took five more years of costly, lost appeals until *RCA* finally was forced to accept the inevitable and settle the case with Farnsworth.

Farnsworth was paid one million dollars over a ten-year period, plus license payments for his patents—a considerable improvement over the original 100,000-dollar offer made by Sarnoff.

Although this seemed like a major win for Farnsworth at the time, fate was not on his side: Japan's attack on Pearl Harbor forced the United Stated to join the Second World War, and this put all development in the area of television broadcasting into the deep freeze for the next six years.

After the war, Farnsworth's patents expired just before the television hit its hockey stick curve: to the major loss of Farnsworth, the promised license payments never materialized.

The fixed payment, however, remained, and with his newly acquired financial independence, Farnsworth went on to study a multitude of different subjects, ranging from nuclear fusion to milk sterilization systems. He also patented the idea of *circular sweep radar display*, which is still the conceptual basis of modern air traffic control systems.

Despite having filed some 300 patents during his lifetime, Farnsworth did not ever succeed on the same scale as with his television invention, finally losing his wealth and falling into bouts of depression and alcoholism. The same personal issues had already been apparent during the stressful litigation process by *RCA*.

Philo T. Farnsworth died of pneumonia at the age of sixty-four.

Later in the same year of 1971, his major opponent David Sarnoff also died, at the age of eighty.

It's hard to determine whether David Sarnoff was a true genius, or just a ruthless businessman who over and over again happened to be in the right place at the right time. But it is clear that the relentless litigation from *RCA* against Farnsworth was the prime cause of Farnsworth's mental and physical breakdown later in his life.

Although some sources even dispute the existence of the original *Radio Music Box Memo*, referred to in Chapter 4: *The Golden Age of Wireless*, there is no doubt that *RCA* became the first true broadcasting conglomerate under Sarnoff's lead. But he was not shy of trying to wipe out Farnsworth from the pages of history: the 1956 *RCA* documentary *The Story of Television* makes no mention of Farnsworth—the storyline has "General" Sarnoff and Zworykin somewhat clumsily praising each other as the two masterminds behind television and citing only *RCA's* achievements as "history firsts". According to *RCA*, the television age started from the speech and subsequent demonstrations that Sarnoff made in the *New York World's Fair* in 1939, although Farnsworth's experimental television system had been on the air five years earlier.

This kind of creative storytelling is a prime example of the power that big companies with well-funded Public Relations (PR) departments may have in rewriting history to their liking.

But in the 1930s, David Sarnoff, having just experienced the radio broadcasting boom, clearly understood the huge business potential of television and was determined to get the biggest possible slice of it. *RCA* had made a formidable war chest from the radio broadcasting business, and Sarnoff wanted to ensure similar success with this new medium.

Although the basic principle of various video camera tube solutions that started emerging in the 1930 and 1940s followed Farnsworth's approach, there was no agreed method for converting the image information into electric signals—the actual implementation of video transmission methods across the early solutions created around the world varied wildly. Even in the United States, *RCA* initially used different standards for different geographical areas, which in practice wiped out any potential for large-scale deployment of television.

To sort this out, *RCA* ended up spending over 50 million dollars to perfect the image capture, transmission and display processes and standards for black-and-white television only. This was an enormous sum at the time—slightly more than the cost of the Empire State Building in New York, or about twice the cost of the Golden Gate Bridge in San Francisco.

But this work was essential: in order to become a nationwide success in the United States, television needed a common standard, and in 1941, a 525-line transmission model was adopted, just before the Pearl Harbor attack stopped all further television activities.

After the war, with a common standard in place, the manufacturers were able to start producing compatible television receivers with ever-lowering prices, opening the way to explosive consumer growth that was similar to what had happened with broadcast radio some twenty years earlier. The concept of broadcasting remained the same, but the end-user experience was now immensely improved by the added power of moving imagery.

The enhancement work that *RCA* made in the field of video camera tubes has an interesting twist: before the litigation process had started, Farnsworth gave Zworykin a step-by-step walkthrough of his image dissector tube creation process, in the hope of getting *RCA* to finance his research. Instead, Zworykin send a detailed description of the image dissector manufacturing process back to *RCA*'s offices in a telegram, so that when he returned to the laboratory, a copy of Farnsworth's image dissector was waiting for him.

Therefore, even though the ongoing patent litigations were still in full swing, the vast resources of *RCA*'s laboratories were busy at work, now helped with the information that the competitor had voluntarily presented to Vladimir Zworykin. A stream of new generation video camera tubes flowed from the laboratories, each sharper and more sensitive than the previous one. Yet there was still plenty of room for improvement: the very early versions had needed immensely bright lights in the studio, and the presenters were forced to wear green and brown face makeup as the video camera tubes had problems with intensive colors like red lips and bright white skin.

And the presenters were originally all white.

The first African-American news anchor, Max Robinson, had already been the invisible voice of television news for ten years before getting his place under the bright spotlights of a live studio. He had actually deliberately shown himself on the air in 1959 and was fired the next day as a result, only to be immediately hired by another station. Eventually he got his spot in front of a camera, and became a very successful news anchor for the *Eyewitness News* team in 1969.

On the other side of the Atlantic, the *BBC* had been busy transmitting TV broadcasts in central London since 1929, the same year that Farnsworth managed to get his first all-electronic image dissector up and running. But the British system was based on a mechanical setup, created by a Scotsman John Logie Baird, who had demonstrated a functioning Nipkow disk-based apparatus in 1926.

The mechanical system had a severely limited resolution: Baird's first demonstrator only supported five scan lines for the whole image. After studying the resolution required to show a distinguishable human face, Baird changed the number of lines to thirty. Over the years he improved the scanning mechanism and managed to reach a resolution of 240 lines, which was very high for a mechanical system.

Baird was very prolific with his television experiments: he kept on adapting the system to cover many new use cases, succeeding in sending long-distance transmissions via phone line from London to Glasgow and later even to New York. All this progress was very impressive, considering that it happened around the same time as Farnsworth was developing the very first fully electronic version of his image dissector.

But with heavy, fast spinning disks, these systems were clumsy and noisy, and worst of all, very limited in terms of active focus depth and acceptable light level: during the *BBC* test transmissions, the presenters had to do their performance in an area of about 0.5 × 0.5 m in order to produce a sharp image.

In his early public transmission experiments, Baird was using the nightly downtime of a *BBC* radio transmitter, but this transmitter could send only video or sound at one time—therefore the transmissions had alternate segments of mute video and blank image with sound, which made it very awkward. It was like the silent films, except that the text frames were replaced by real audio. Eventually the *BBC* became interested enough in his work to provide him with two dedicated transmitters: one for video and one for sound. Hence, Baird became the first person in the world to provide actual television broadcasts with simultaneous video and sound.

An interesting anecdote about Baird and Farnsworth is that these two pioneers actually met in 1932: Farnsworth was looking to raise money to cover the cost of the *RCA* litigation by selling a license to Baird, not knowing that Baird actually was not a very wealthy man. During this meeting they both made demonstrations of their respective systems, leaving Baird really worried: what he was shown by Farnsworth appeared to be miles ahead of his mechanical television solution, and to be able to ensure a positive outcome to his future work in the face of impending defeat in the system wars, he suggested a cross-licensing deal. Although this was not financially what Farnsworth came to London for, he accepted.

With the continuous advances in the field of electronic television, the writing was on the wall for Baird's mechanical solution: a British joint venture, the *Marconi-EMI Television Company*, had acquired rights to *RCA's* picture tube technology, and invited the experts from the *BBC* to see a demonstration of their fully electronic television. As a result, the *BBC* started running the Baird and the *Marconi-EMI* systems on alternate weeks through its transmitters. The idea was to get fair customer feedback on the comparative quality of these systems, although by just looking at the specifications, there was no true competition: 240 scan lines per frame and severely restricted studio-only focus simply could not compete against *Marconi-EMI's* crisp 405 scan lines with normal, film-like depth perception. The feedback came not only from the viewers, but also from the performers, who begged the producer not to assign them to be on the air during the "Baird weeks".

The duality of the *BBC's* transmissions forced Baird to adapt his commercial receivers, creating the world's first dual-standard television *Baird Model T5*, which could display both his and *Marconi-EMI's* transmissions.

When the electronic television was chosen as the system of choice also in Germany, it was evident that the future of television would not have space for mechanical solutions. Still, to commemorate the father of the first patented method of television, the first German television station was called *Fernsehsender Paul Nipkow*.

The *BBC* switched to *EMI* electronic video cameras in 1936, the quality of which kept on improving: in May of 1937, a version called *Super-Emitron* was sensitive enough to provide the first outdoor transmission from the coronation of King George VI.

Although the loss in the systems fight initially frustrated Baird, he understood that the *Marconi-EMI* system had truly won on merit, and moved his own experiments into the electronic domain as well, thanks to the cross-licensing deal with Farnsworth. Fate also stepped in: Baird's laboratory, with his latest mechanical prototypes, was destroyed in the fire at London's iconic Crystal Palace on November 30th, 1936, making it easier to start with a clean slate. The results of the fire were not as bad as they could have been, as Baird had an insurance policy covering his equipment at Crystal Palace.

Baird kept on innovating in this area, creating the first functional demonstration for color television in 1944, and even patenting his 500-line 3D-television prototype.

After the war, Baird presented a grand plan for a 1000-scan line color television, *Telechrome*, managing to get the *BBC* initially interested in funding the creation of such a system.

In terms of picture quality, *Telechrome* would have been almost on par with today's *digital High Definition (HD)* television systems, but as an analog system, it would have needed a massive amount of bandwidth per channel. This, in turn, would have limited the total number of television channels available, for reasons which are explained in TechTalk *There is No Free Lunch*.

Unfortunately, the ongoing post-war effort in Britain put a major damper on available resources, and *Marconi-EMI's* 405-line version remained in place.

All television transmissions had been halted for the duration of the Second World War, but when the *BBC* restarted the transmissions in 1946, television in Britain hit the hockey stick curve: 20,000 sets were sold in the first six months alone, which is a lot for something that was still considered as a luxury item after the war.

At the same time in the United States, commercial television had been off the air during the war, as all prominent electronics manufacturers participated in the war effort.

Ironically, Philo T. Farnsworth's *Farnsworth Television & Radio Corporation* had its best financial run while manufacturing radio equipment for the U.S. Army during the war, including a high-frequency transceiver with type designation *BC-342-N*, which was a 115-volt version of a widely used field and aviation radio *BC-342*. But despite the ample funding provided by this lucrative wartime contract, *Farnsworth Television & Radio Corporation* went out of business just before the post-war television manufacturing hockey stick curve started in the United States.

Farnsworth's name lived on after the war, though: thanks to their reliability and ease of use, thousands of surplus *BC-342-N* radios by *Farnsworth Television & Radio Corporation* found a new life as Radio Amateur equipment.

Although there were good technical reasons why the whole world could not use the same television standard, national protectionism also had its share in this game. Yet, the basic concept of capturing and displaying images based on stacked scan lines was the same in all of them—only details like the number of scan lines and the number of frames per second were different, and thanks to this, the makers of the television sets were able to isolate the necessary bits that made up the regional differences, paving the way for a true mass-market production—most of the components needed for a television set remained the same, independent of the target market.

Later down the line, some television manufacturers followed the *Baird Model T5*-approach, dynamically supporting multiple standards, which was a boon for people living near to the boundaries of regions that had adopted different broadcasting formats.

On the technical side, the reason for having different quantities of scan lines and frame rates in different countries stemmed from the frequency used for the distribution of electric power. For example, Europe uses 50 Hz Alternating Current (AC) power, whereas the United States and many other countries have 60 Hz as the standard.

The human eye is not fast enough to detect the fact that certain lamp types actually dim and brighten up in sync with the power line frequency, effectively blinking continuously with high speed, incandescent lights being some of the worst to exhibit this phenomenon. But the scan speed of television cameras is fast enough to detect this, and thus has to be synchronized with the power distribution frequency—if there was a mismatch, moving bands of darker and brighter areas would appear on the television screen due to the *stroboscopic effect*. This could be avoided by selecting a frame rate of 25 frames per second for countries with 50 Hz AC power frequency and 30 frames per second for countries using 60 Hz AC power frequency: although the sampling would occur only by half of the power frequency, the samples would always hit the same phase of the ongoing power cycle.

The resulting frame rate, together with the available bandwidth of the transmission channel, assert further mathematical limitations on the number of scan lines that can be embedded inside a single image frame.

Other variables that added to the plethora of television standards that were used in different geographical areas included the *frequency separation* of audio and video portions in the transmission signal, and eventually, the method used for color information encoding, which was a bolt-on enhancement to existing black-and-white television transmissions.

As a side note, there really was no reason to have different frequencies for AC distribution—it all boiled down to protectionism between markets: at the beginning of the 20th century, the United States, with the suggestion by Tesla, chose 60 Hz, whereas Britain went with the German 50 Hz version, both effectively making it more difficult to import each other's electric motor-based devices. As a result, the rest of the world was simply divided according to which side of the Atlantic happened to have a greater local influence at the time of rolling out the local electric power networks.

In Britain, the next post-war technology step happened when the *BBC* switched to the 625-line Western European base standard in 1964. Yet, despite sharing a common European standard, the United Kingdom still chose a different frequency separation between audio and video portions within the television transmission signal.

I had a somewhat humorous personal experience based on this minor incompatibility when I moved from England to Finland and brought my British TV set along: an electrician came to install a television antenna on the roof of my house while I was away, and he spent hours re-tuning and tweaking the system, trying to figure out why he was getting a perfect picture but no sound, assuming that the antenna amplifier would somehow be the culprit.

He was relieved when I arrived home and explained the reason for this mysterious behavior. The actual problem was easily circumvented by using my multi-standard video recorder as a front-end tuner, delegating the television to a monitor mode through the *Syndicat des Constructeurs d'Appareils Radiorécepteurs et Téléviseurs (SCART)* connector: the chunky European standard connector for high-quality analog video.

When dealing with fundamental systemic upgrades in an established mass-market environment, one important aspect is the drive for *backwards compatibility* whenever possible: a good example of how this could be handled was the way stereophonic sound was added to FM radio, as discussed in Chapter 4: *The Golden Age of Wireless*.

The introduction of color into television transmissions had to be provided in a similar manner, so that the television sets that were already out in the market would not become obsolete.

The existing scan lines in the black-and-white television signal essentially already contained the brightness, or *luminance* information, of each line, so all that was needed was to get the color information, or *chrominance* information somehow added into the existing structure of a video signal.

This was done by embedding a high-frequency *color burst* into the transmitted signal, timed to happen between the transmission of two scan lines. This was a period during which black-and-white televisions were not expecting any information, thus effectively ignoring it. But the new color television receivers were able to extract this additional chrominance information, join it with the luminance information of the received scan line, and then proceed to display the scan line in full color.

As the luminance and chrominance information for a single scan line did not arrive exactly at the same time, the whole process of resynchronizing them was quite tricky to implement with the analog circuitry of the time, but this solution provided full interoperability with the millions of black-and-white television sets that already existed.

Both the addition of stereophonic sound and color video are great examples of a solution in a situation where an existing, widely adopted system cannot be fundamentally changed—the human ingenuity beautifully steps in to create a fix that is backwards compatible and does not disrupt the existing customer base.

As for the analog color systems, the United States was first with the idea of an embedded color burst as an extension to their 525-line black-and-white *National Television System Committee* (*NTSC*) solution. To accommodate this additional data burst so that it did not cause interference with the luminance information, the frame rate in the United States had to be slightly tweaked from 30 frames per second to 29.97 frames per second.

Unfortunately, the straightforward, first-generation color encoding solution used in this system had a tendency to distort colors in certain receiving conditions, and the NTSC acronym jokingly got a new meaning: "*Never the Same Color*".

The color distortion was mainly caused by *multipath propagation interference*, in which the received signal reaches the antenna both directly and via a reflection from a nearby object.

This inherent limitation of the NTSC system led to the creation of the *Phase Alternating Line* (*PAL*) standard, which, in the late 1960s, became the standard in Western Europe as well as in Brazil, although Brazil had to use a slightly modified, 30 frames per second version due to their 60 Hz power line frequency. As the name of the standard describes, by alternating the color pulse phase between scan lines, it effectively evens out any phase errors in the received signal, providing a consistent color quality.

A third major system, *Séquentiel couleur à mémoire* (*SECAM*) was developed in France. It was taken into use in France and many of the former colonies and external territories of France. SECAM was also chosen by the Soviet Union and forced upon most of the Cold War Soviet Bloc countries, not due to technical merits, but to reduce the "bad western influence" that might be caused by watching western TV transmissions.

As a result, along the borders of the Iron Curtain, you could still see the neighboring TV transmissions, but unless your TV was a multi-standard version, you were limited to watching a black-and-white image without sound. Therefore, it was no surprise that one of the best-selling electronic upgrades in East Germany was a PAL-to-SECAM conversion kit for television, as somehow the West German programming was perceived to be more relevant than the East German one, especially when it came to news.

In the end, even the television sets made in East Germany had PAL-compatibility built in.

This politically driven, artificial East–West technological divide, however, was not entirely complete across the regions: Albania and Romania had PAL in place, and outside of the Soviet dominance, Greece had also originally selected SECAM, although they migrated to PAL in 1992.

It is hard to exaggerate the propaganda value of television: daily access to this "bad western influence" probably played a big part in the crumbling of the Cold War divide that led to the fall of the Berlin Wall and the reunification of Germany.

The introduction of color television added to the potpourri of standard combinations in use, so when the *International Telecommunication Union* officially defined the approved set of analog television standards in 1961, there were fifteen slightly different versions to choose from.

When the world eventually migrated to digital terrestrial television, we still ended up with four different terrestrial transmission standards: Japan, Europe, United States and South America all decided on different formats, based partly on their technical merits, partly on pure protectionism again.

Yet we have gone down from fifteen terrestrial standards to four, so there has been a clear improvement in the past 50 years.

Concepts like *analog* and *digital* are discussed in TechTalk *Size Matters*.

All this analog mix-and-match is now slowly but surely vanishing into history with the global deployment of digital television. The predecessor for this step up in image quality was *Sony's* 1125-line analog *High Definition Video System (HDVS)*, which was introduced in Japan at the end of the 1980s.

Thanks to cheap, ultra-fast digital signal processing and improvements in computing power, it is now possible to compress both video and audio content into a digital data stream which requires much less bandwidth than what would be needed for transmitting the same content in analog format. The issue with any analog television transmission is that it is constantly hogging the full available channel bandwidth, whether it is showing just a black screen or a scene packed with action.

In contrast, digital television transmission structure relies on the existence of local image memory in the receiver, and in essence only transmits fine-grained differences between individual frames. This means that the amount of bandwidth required to transmit video at any given time varies greatly depending on the ongoing content. The maximum available bandwidth dictates the upper limit for the picture quality during the most demanding scene changes: with lots of movement, like during explosions or aggressive camera motions, noticeable *pixelation errors* may therefore become visible for a very brief moment.

Another major enhancement stemming from the extreme malleability of a digitally formatted data is the possibility of not only selecting the desired resolution of the channel, but also bundling several digital channels together into one data stream. This allows one physical television transmitter to broadcast multiple channels simultaneously inside a single broadcast signal, only to be finally extracted into separate channels at the receiving end. This was briefly mentioned already in the discussion about Digital Audio Broadcast (DAB) in Chapter 4: *The Golden Age of Wireless*.

As a result, if you take advantage of the lower required bandwidth for digital channels and bundle up four channels, you can turn off three physical transmitters and still enjoy the same number of channels as before. The fact that all these four channels actually come through a single transmission stream is totally transparent to the end users.

This brings along several benefits:

First of all, a considerable amount of electricity can be saved, as television transmitters tend to need power in the range of tens or even hundreds of kilowatts and are often running 24/7.

Secondly, maintaining the transmitters and antenna towers is costly, and, by bundling the channels into one actual broadcast transmission, a lot of this hardware can be decommissioned.

But most importantly, this method saves the precious radio frequency spectrum, which can then be reused for other purposes. This kind of freed spectrum is actually extremely valuable wireless real estate, and the benefits of reusing it are discussed further in Chapter 10: *Internet in Your Pocket*.

If we continue with the example above and only decommission two transmitters, changing the remaining two into four-channel digital ones in the process, we have not only saved a lot of bandwidth and running costs, but as a bonus, now have twice the number of channels at our disposal. These extra channels can then be leased to the highest bidder, providing more potential income for the network providers.

More discussion of bandwidth issues can be found in TechTalk *There is No Free Lunch*.

An additional benefit of switching into the digital domain comes from the fact that both audio and video quality remain much more consistent when compared with traditional analog transmissions: in good enough reception conditions, the only limiting factor for the video and audio quality is the maximum allocated bandwidth on the channel.

The negative effect that follows from digitalization is the fact that any errors in reception are usually much more annoying than was the case with analog television: as the digital video signal is heavily based on only managing the differences between frames, any gaps in the reception can leave big chunks of the image distorted or frozen for a couple of seconds, and cause nasty, screeching and metallic-sounding artifacts in the audio part of the transmission. These issues are most pronounced in mobile reception conditions—with fixed terrestrial television setups and cable or satellite television, as long as the received signal strength is above a certain threshold, the received picture and audio quality remains as good as it can theoretically be.

The flipside of switching into digital television is that there is no way to remain compatible with the old analog television transmissions, so this is a true generation change. All new digital television receivers still have circuitry to handle analog transmissions, but as the transmitters are switched from analog to digital mode, old analog television receivers are eventually going to go the way of the dodo.

The enhancements brought by digital television's better quality and lower required bandwidth were first taken into use in satellite television at the turn of the 21st century. The savings gained from cramming the maximum number of channels into one transmitted microwave beam from the satellite were significant, and the high quality achieved by digital transmissions was paramount for service providers that were competing against cable television.

To receive satellite television, a special receiver is always needed, so it was possible to add the necessary electronics for converting the digital signal into an analog one, making it easy for the consumers to keep using their existing television sets in monitor mode.

The main driver of the digitization of satellite television was the cost of setting up the required space infrastructure: the lighter you can make a satellite, the cheaper it is to launch into space, so you try to minimize the number of individual, heavy microwave antennas and the required parallel transmitters. Less electronics means fewer items that can break during the expected lifespan of the satellite, and a reduced number of beams means that lighter solar panels can be used to power the satellite, again reducing the overall weight and complexity.

Because the launch and actual development of a space-grade satellite were by far the costliest items in setting up a satellite distribution system, adding some expensive, leading-edge digital circuitry to reduce the number of beams actually meant that the overall cost of a satellite launch was reduced.

On the receiving side, customers tend to have multi-year relationships with their satellite television providers, so the additional cost caused by the digitalization of the satellite receivers could be reaped back over the years as part of the monthly subscription.

You can read more about satellite television systems in Chapter 7: *Traffic Jam over the Equator*.

Although cable television systems are not wireless, they have been an essential part of the broadcast television revolution, and hence are worth a sentence or two in this context:

Most urban areas offer cable television systems for transmitting hundreds of channels, effectively creating an isolated version of the radio spectrum inside the cables of their distribution network. The extra benefit of the shielded cable environment is that there are no other transmitters or regulatory limits to worry about:

all those frequencies in this tiny, completely isolated universe inside the cable belong to the cable company, and they can divide and conquer them as they please.

This setup can even be utilized for two-way traffic, which is now widely used by the cable companies to provide Internet and fixed phone line services in addition to traditional cable television programming.

As the number of available cable channels can now be calculated in the hundreds, going digital was the only way to cater for the ever-increasing demand for capacity and the quest for improved picture quality. As with satellite providers, this transition to digital delivery was helped by the fact that many cable companies still deliver their subscription-based programming through separate set-top boxes, which can then convert the signal to analog format: although the incoming programming stream is digital, customers do not need to upgrade their television sets.

The latest approach to television content delivery is to go fully digital, using the Internet as the delivery medium: instead of having all the channels stacked up inside the distribution cables in parallel, the customer picks up the digitized video stream of the channel of interest, and the content is then delivered like any other Internet-based data.

The use of the Internet has also opened the formerly very expensive-to-enter market of broadcasting to entirely new *on-demand* providers, like *Netflix*, *Amazon* and *Hulu*, and these new players now threaten both the vast satellite television business and the traditional cable television business. Those incumbent providers that are not already active within the massively growing Internet-based delivery approach will most likely face major problems with this latest technological shift, and therefore it is no surprise that hybrid solutions are already being announced by proactive companies. As an example, the prominent satellite television provider *Sky Plc* has announced that its complete subscription television package will be available via the Internet across Europe in 2018.

Thanks to the new access mode offered by the Internet, even the good old video recorders are going virtual. The content of every channel is constantly recorded at the network provider's digital storage system, and the customers can go back in time at will, accessing the digitized media from a shared, single storage.

This kind of *time shift-capability*, together with the rise of on-demand providers, is fracturing the traditional television consumption model. Apart from certain media content that has clear and inherent immediate value, like live sports events or breaking news bulletins, customers can now freely choose the time, and increasingly also the place of their media consumption.

Time shifting also makes it possible to skip advertisements, which poses a major risk for providers of non-subscription services, whereas subscription-based services are booming as customers are happy to cut out annoying advertisement breaks from their media consumption time in exchange of a few dollars per month.

But whatever the method by which the programming is delivered, television has totally revolutionized the way people spend their waking hours. The time spent watching television has grown year after year, and is currently a major chunk of our lives: in 2014, the average American, the current record holder, spent 4.7 hours daily watching TV. If you take off time spent sleeping, eating and working, not much is left for other activities. Only recently the Internet has finally risen as the new contender for eyeball time.

The lure of television stems from the fact that humans are visual beings: up to 40% of our brain's *cerebral cortex* surface area is dedicated to vision.

Because our survival used to depend on detecting potential predators before they came close enough to make a meal of us, evolution optimized our perceptive skills to be attracted to even the slightest movement in our field of vision. Hence a switched-on television set is constantly demanding our attention. This is a "built-in" feature in our brains, and thus very hard to resist.

Due to its addictive nature, television was by far the most significant change brought into our lives by the application of radio waves—until mobile phones came around.

As technology advanced, we have been able to cram ever more information into the limited radio spectrum that is available to us. Thanks to the advances in solid-state electronics, sensor elements for video cameras no longer need a vacuum and an electron beam to extract the image frames. Likewise, on the receiving side, bulky CRT screens have become relics of our analog past, replaced by *Light-Emitting Diode* (*LED*) and *Liquid Crystal Display* (*LCD*) based flat screens.

None of this recent progress takes away the fact that everything you see on television only became possible through the hard work and ingenuity of Philo T. Farnsworth and his peers, who built their solutions on top of the inventions and theories of Hertz, Maxwell, Marconi, Tesla and so many others.

This ability to constantly improve and expand the work of earlier generations is one of the most prolific aspects of humanity, and it still drives the utilization of the invisible waves that were first harnessed just over one hundred years ago.

But as Tesla demonstrated in the very early days of the wireless revolution with his remote-controlled prototype, these invisible waves can also be used for entirely other purposes than just transmitting audio and video to consumers.

One of the most prominent of these alternative uses that most of us are totally oblivious of is enabling millions of air travelers to get safely from A to B every day, in almost any kind of weather. Let's take a look at these *Highways in the Sky* next.

# Chapter 6
# Highways in the Sky

By 1937, Amelia Earhart, "the Queen of the Air" according to the media of the era, already had several historical firsts in her portfolio of achievements: she had been the first woman to fly solo over the Atlantic, not only once but twice, and she also held the altitude record for female pilots.

As the next step, Earhart had just recently attempted to be the first woman to fly around the world, only being immediately forced to abort her attempt due to technical issues with her airplane. But the moment the plane was repaired and ready to fly again, she restarted her attempt, this time from Oakland, California.

On July 2nd, after about a month and a half of flying and with 29 intermediate stops behind her, she had loaded her shiny *Lockheed Electra* twin engine plane up to the hilt with gasoline on a small airfield in Lae, New Guinea, and was ready to embark on the most dangerous part of her circumnavigation.

Earhart had only three legs to go until she would be back in Oakland.

Also on the plane was her sole crew member Fred Noonan, a distinguished navigator, who had helped her on her arduous eastward travel around the globe.

It had taken them 42 days to reach their current stop at Lae, and the next flight was expected to cross the first stretch of the vast emptiness of the Pacific Ocean—from Lae to the tiny, two kilometer-long and half a kilometer-wide piece of coral called Howland Island, over 4,000 kilometers and 18 flight hours away.

The size of the Pacific Ocean is hard to grasp: it covers almost a third of Earth's surface, and when seen from space, it fills almost the entire visible globe, with only a couple of tiny specs of land breaking the blue vastness of the water.

Even today, many of the islands in the Pacific don't have airports and get a visiting ship only once per month or even more rarely, so if you are looking for a place away from the hustle and bustle of the modern world, some of the remote Pacific islands should definitely be on your list.

In 1937, the existing navigation methods for crossing oceans were based on the naval tradition that had been in use for about 200 years, since John Harrison solved the problem of finding your longitude: measurements of the positions of the Sun,

© Springer International Publishing AG, part of Springer Nature 2018
P. Launiainen, *A Brief History of Everything Wireless*,
https://doi.org/10.1007/978-3-319-78910-1_6

Moon or stars are taken with a sextant and an accurate clock to get a fix on the current location.

Based on these observations, you calculated your expected compass heading, trying to compensate for the prevailing winds as well as possible, by checking the direction of waves and relying on the sparse weather reports available for the area.

Airplanes flying across vast stretches of ocean used exactly the same approach, so it was no surprise that Fred Noonan was not just a navigator but also a certified ship captain. He had recently participated in the creation of several aviation routes for the seaplanes of *Pan American Airways* across the Pacific, so tackling a challenge like this was not unfamiliar to him.

But *Electra* wasn't a seaplane, so it had to find a runway at the end of the flight, and Howland Island was the only one within range that had one.

To hit such a small target after such a long flight was very ambitious because there were only a couple of other islands along the way that could be used to verify their actual position. Thus, to alleviate this problem, the plan was to get some extra help from novel radio location technology on board U.S. Coast Guard ship *Itasca*, which was anchored at the island—*Itasca's* radio transmitter was to be used to provide a locator beam, guiding *Electra* to its destination.

*Electra* had a brand-new *direction-finding receiver* on board, which had been installed at Lae just prior to this flight, and this method of medium-range navigation was expected to provide a sufficient backup against any navigational errors that were caused by flying for hours over open sea.

Similarly, *Itasca* also had direction-finding equipment on board, which could be used to get a bearing on the incoming plane's transmissions, and any corrective guidance could be relayed to Earhart.

It was inevitable that after such a long flight with only a handful of physical locations for getting a reliable location fix, ending up *exactly* at Howland Island after 18 hours of flight time would be an unlikely outcome. Therefore, the help of radio navigation would be most welcome.

The idea of the *Radio Direction Finder (RDF)* system is simple: a fixed radio transmitter sends a low-frequency signal at a known point on Earth, and the receiver uses a rotating loop antenna to determine the direction of the incoming signal. At a perpendicular position to the loop, the signals received by the edges of the loop cancel each other out, and a distinctive drop in signal strength is observed. By reading the angle of the antenna at this "null position" it is possible to determine the relative bearing of the transmitting station and align the plane's heading accordingly.

The RDF system does not even require a dedicated transmitter—any radio transmitter that sends on a compatible frequency will do. Therefore, as RDF receivers became more commonplace in planes after the Second World War, strong broadcast transmitters were often used for this purpose, with the additional benefit of providing music and other entertainment for the flight crew.

But in 1937, in the middle of the Pacific, the only transmitter to be found was the one on board the *USCG Itasca*, and as expected, around the expected time of arrival, *Itasca* received ever-strengthening audio transmissions from Earhart.

She reported poor visibility due to overcast clouds and asked the *Itasca* crew to figure out the bearing of *Electra* by the audio communications she made, apparently not realizing that the RDF receiver on *Itasca* did not support the active audio transmission frequency used by Earhart—she clearly was not really familiar with this new technology. Although it was obvious from the strength of the very last received signals that the plane was indeed somewhere near to the intended destination, no two-way communication was ever achieved, for reasons unknown. Earhart even acknowledged the reception of *Itasca's* RDF beacon signal at her end, but indicated that she could not get a bearing from it, without giving any details as to why.

The last received transmission at 8:43 AM stated the assumed position and direction of the flight and that they were flying at a very low altitude, low on fuel, trying to spot any land. As a last resort, *Itasca* used the ship's boilers to generate a huge plume of smoke for Earhart to see, but there were no further transmissions received from the plane.

*Electra* disappeared, seemingly without a trace.

The U.S. Navy did an extensive search in the nearby areas but did not find any sign of the plane or the crew.

The final location of *Electra* and the destiny of its crew remains an unresolved mystery to this day, and the search for the remains of the crew and the hull of *Electra* still continues to make headlines at various intervals. Findings of human bones on nearby atolls have been reported, but at the time of writing this book, no confirmation of a definite match with either Earhart or Noonan has been announced.

Both crew members were officially declared dead two years after their disappearance.

The investigation into the reasons behind the failure was able to indicate several potential problems:

First, the weather was poor, and distinguishing low cumulus cloud shadows from the profile of a small island while flying can be very difficult in such a situation.

Second, the assumed position of Howland Island was found to be off by about 9 km. This is not much in aviation terms, but it could have led Noonan off course just enough during the final moments of the approach to make his target island vanish in the haze that was caused by the poor visibility. With this level of error, they could have flown past their destination, and then ended up circling over the open water surrounding Howland Island.

Perhaps the biggest mistake Earhart made, though, was the fact that she did not practice the use of her newly installed RDF equipment. She had only been given a brief introduction to the system prior to takeoff at Lae.

This may sound like deliberately asking for trouble, but from her point of view, the RDF system was only a backup measure: they had already crossed two thirds of the globe by means of traditional navigation methods, and Earhart had probably gained a false sense of confidence due to her earlier successes along the way.

Unfortunately, the leg from Lae to Howland Island was the most complicated and unforgiving part of the journey, with no room for error, and it is easy to say in hindsight that spending a bit more time familiarizing herself with this new technology could have turned a tragedy into a triumph.

This leads us to the crucial point, applicable to any of the latest, shiniest navigation equipment at our disposal: even the best equipment is worthless unless you know how to use it, and, in case of malfunction, you should always maintain a general sense of location so that you can proceed to your destination by using other means of navigation.

Always have a backup plan.

There are plenty of recent horror stories of persons blindly following the guidance of their navigation equipment, leading them sometimes hundreds of kilometers off the intended destination, or even into life-threatening situations, simply due to inaccurate mapping data or incorrect destination selection.

If this is possible with the extremely easy-to-use equipment we currently have, it is not difficult to see how Earhart could have made an error with her much more primitive setup.

RDF systems had gone through several evolutionary steps prior to ending up as auxiliary equipment on the *Electra*. The accuracy of such a system was partly dependent on the size of the loop antenna, which again was limited by the physical measures of an airplane and the extra drag an external antenna generates during flight.

Due to this, the very early systems used a reverse configuration, in which the transmitting loop antenna was rotated with a fixed speed, synchronized to a clock, and the receivers could determine their bearing from the time it took to detect the lowest level of signal.

As discussed in Chapter 3: **Radio at War**, this type of technology was first used to guide the German *Zeppelin* airships, between 1907 and 1918, most notably during the First World War bombing raids on London.

RDF systems are quite limited in their usefulness, and improvements in electronics finally led to the current, ubiquitous *VHF Omni Range* (*VOR*) systems, which, instead of sending just a homing beacon, make it possible to determine on which particular *radial* from the VOR transmitter you currently reside.

The radial information is sent with one-degree resolution, embedded by altering the phase of the signal in relation to the transmission angle, so no mechanical movement is necessary either on the transmitter or the receiver side. The receiver only has to extract the phase information from the received signal in order to detect the current radial related to the fixed location of the VOR transmitter.

You can accurately pinpoint your position by getting radial fixes from two separate stations and drawing the corresponding lines on your aviation chart. The position on the map where these two radials cross is your current location.

Some VOR stations even contain an additional *Distance Measurement Equipment* (*DME*) setup, which allows airplanes to figure out their *slant distance*, which is the direct line distance from the airplane to the ground antenna. This allows you to get an accurate position fix from only a single VOR station.

DME is based on the same principle as *secondary radar*, which will be discussed in Chapter 12: **"Please Identify Yourself"**, except in this case, the roles of the airplane and the *transponder* are reversed—the airplane does the polling, and the ground station responds to the query signal. The airborne receiver can then determine the slant distance by calculating the elapsed time between query and response.

The origins of DME date back to the early 1950s, when James Gerrand created the first functional system at the Division of Radiophysics of the *Commonwealth Scientific and Industrial Research Organisation (CSIRO)* in Australia. DME became an International Civil Aviation Organization (ICAO) standard, and it shares the same principle as the military version, *Tactical Air Navigation System (TACAN)*.

VOR technology was first used in the United States in 1946, and it remains the backbone of current aviation navigation: *airways* crisscross the skies all around the world, based on virtual lines between *waypoints*, such as airports, VOR stations and known intersections between two fixed radials. Flight plans are constructed as sets of these waypoints, leading from one airport to another in a pre-determined fashion. These airways create highways in the sky, shared by multiple planes, separated either by altitude or horizontal distance by *Air Traffic Control (ATC)*.

For the final approach to major airports, another radio system called *Instrument Landing System (ILS)* is used, and it dates back even further than VOR: the tests for an ILS system started in the United States in 1929, and the first working setup was activated at Berlin Tempelhof airport in 1932.

ILS creates a "virtual" *glideslope*, a highly directional radio signal with a small, usually three-degree vertical angle, which can be used as a guide to the beginning of the runway, both horizontally and vertically. The signal can be followed either by manually looking at the *glideslope indicator*, or automatically by an autopilot that follows the invisible electronic "downhill" towards the runway.

With this system, planes can approach the airport with zero visibility, with the aim of ending close enough to the runway to be able to finally see the runway lights before a specified *decision altitude* is reached. If there is no visual contact at this point, the plane has to perform a *missed approach*—pull up and try again by returning to the initial acquiring point of the glideslope, or give up and fly to an alternative nearby airport that reports better weather conditions.

With *Category IIIc ILS*, an autopilot with matching *autoland* capabilities can perform a complete landing without any interaction from the pilot, and in zero visibility. The first such landing happened in Bedford, England in 1964. Most modern airliners have the autoland feature in their autopilot system, and to maintain certification, it must be used at least once per 30 days, whatever the weather, so chances are that those of us who fly frequently have experienced an automatic landing at some point.

The newest radio navigation aids that have taken the world by a storm are satellite-based navigation systems.

Unfortunately, this technology was not yet available on the evening of September 1st, 1983, when *Korean Airlines* flight *KAL 007* was readying for takeoff from Anchorage, Alaska, in order to fly across the Northern Pacific Ocean to Tokyo, Japan:

The transcripts show that the flight plan of *KAL 007* had been accepted by the *Air Traffic Control (ATC)* "as filed", meaning that the requested flight plan that was based on the standard *Romeo 20* airway across the ocean did not need to be modified, and the expectation was that the pilots had entered it accordingly into the plane's *Inertial Navigation System* (*INS*).

In the case of *KAL 007*, this was by no means the first time this standard airway route, Romeo 20, was being flown, and everything should have been entirely routine for the experienced Korean crew.

INS has a map of global coordinates of thousands of *waypoints* in its internal database and is able to guide the autopilot of the plane based solely on the inputs from its internal *gyroscopes* and *accelerometers*, without needing help from other navigation systems. INS simply detects the physical movements of the airplane in all three dimensions, constantly calculating the expected position based on these inputs. This technology was originally developed to aid spacecraft navigation, having its origins in the feared German *V-2* rockets of the Second World War. It was adapted for aviation use in the late 1950s to help navigation in regions with no radio navigation facilities, like during long oceanic crossings around the globe—exactly what *KAL 007* was about to embark on.

The plane took off and was assigned a coarse heading according to the active flight plan, with the request to report back to the ATC when it was passing the BETHEL VOR station, near the Alaskan coastline.

The pilots entered the assigned heading into the autopilot but apparently never switched the autopilot to the INS guidance mode. The assigned initial coarse heading was close enough to the right one, though, and the crew must have been satisfied with the direction they were flying in.

*KAL 007* made the required BETHEL report 20 minutes after takeoff, and the second waypoint in the flight plan, NABIE, was confirmed as the next expected mandatory reporting point. But neither the crew nor the ATC of Anchorage noticed that *KAL 007* had actually flown about 20 kilometers north of BETHEL VOR and thereafter continued flying slightly off course, on a more northerly route. As a result of this wrong heading, *KAL 007* was not going to cross over NABIE.

Despite this, the crew made the expected mandatory radio report, claiming to be over NABIE. It is unclear how the pilots came to the conclusion that they had eventually reached the NABIE waypoint, but they clearly had not been fully aware of their actual navigational situation ever since the very beginning of their flight.

The first indication that something might be wrong was due to the fact that the Anchorage ATC did not respond to their positional radio calls at NABIE, so they relayed their position reporting message via another *Korean Airlines* flight *KAL 015*. That flight had left Anchorage some 15 minutes after *KAL 007* and was listening on the same ATC frequency.

Atmospheric radio interference in high Northern latitudes is not entirely uncommon occurrence, so this glitch with relatively long-distance communications may have happened during their earlier flights as well, and thus was not necessarily seen as a major issue at the time.

But three and a half hours later, due to this undetected course deviation, *KAL 007* was already 300 kilometers off course and entered the airspace of the Soviet Union, over Kamchatka Peninsula. It ended up crossing just north of the city of

Petropavlovsk—a town facing the Northwest coast of the United States. In better visibility, the appearance of city lights along a route that was supposed to fly over open sea would have been an unmistakable warning sign, but *KAL 007* was now flying over a heavy cloud cover that dampened out any light from below.

Most unfortunately, this was all happening at the height of the Cold War, and Petropavlovsk was the home of several top-secret Soviet military installations and airports, aimed at spying on the United States and acting as the first line of defense against any military actions against the Soviet Union.

The recordings of the Soviet Air Defense indicate how this violation of airspace was immediately noticed and several fighter jets were scrambled to respond, soon approaching the plane that was about to enter the international airspace again after crossing over the full width of Kamchatka. It was a dark, cloudy night, and the Soviet fighter pilots failed to recognize the type of the plane, while actively discussing whether it could be a civilian plane.

Even though the pilots could not confirm the type of the plane and had suspicions that it may not be a military target, their controller on the ground instructed them to shoot it down: after all, there had been several earlier spy flights by the U.S. Air Force over the military installations in the area, and the commanders who had failed to respond to these incidents had been fired.

Almost at the same moment as the command to attack was given, the totally oblivious crew of *KAL 007* contacted the *Tokyo Area Control Center* and made a request to climb to flight level 350 in order to reduce fuel burn—a standard procedure when the weight of the plane had reduced enough, thanks to the diminishing amount of kerosene in the tanks.

The altitude change was granted, and as the plane started to climb, its relative horizontal speed was suddenly reduced, causing the tailing Soviet fighters to overtake it. This random change in relative speed was thought to be an evasive maneuver, further convincing the Soviet pilots that the plane really was on a military spying mission and fully aware of being followed.

The jets caught up with the jumbo again, and three minutes after *KAL 007* had made the radio call informing that they had reached the assigned flight level 350, the plane was hit by two *air-to-air missiles*.

Despite losing pressurization and three of its four hydraulic systems, *KAL 007* spiraled down for an additional twelve minutes until it crashed into the sea near Moneron Island.

A total of 269 people, passengers and crew, were killed as a result of a simple pilot error, which could have been avoided by cross-checking their course using any other navigation means that were available in a commercial airliner. There would have been many direct and indirect ways to check the position of *KAL 007*, but apparently the course felt "right enough" to the pilots, and having flown the same route several times in the past, they ended up being oblivious to the mistake they had made.

Just two weeks after this tragic incident, President Ronald Reagan announced that the *Global Positioning System* (*GPS*), a novel satellite-based navigation system that was under construction and had been designed exclusively for the U.S. Military, would be opened for free, civilian use.

The roots of a Global Positioning System go all the way to the very beginning of the *Space Race* between the Soviet Union and the United States: when the very first satellite, *Sputnik*, was launched by the Soviet Union in 1957, it proudly made its presence known to anyone who had a receiver capable of listening to the constant beeping sound at 20.005 MHz frequency.

Two physicists, William Guier and George Weiffenbach, at Johns Hopkins University, studied the frequency shift that was caused by the relative speed of the satellite against their fixed receiver: due to the *Doppler effect*, the frequency you receive appears higher than the expected frequency when the transmitter is moving towards you, and lower when it moves away from you. Based on the dynamic change of this observed shift, Guier and Weiffenbach managed to create a mathematical model that could be used to calculate *Sputnik's* exact orbit.

After some further study, they realized that this process could also be reversed: if you knew the orbit of a satellite and the exact frequency it was supposed to send on, you could use the same mathematical model to calculate your position on Earth based on the dynamically changing frequency shift that you observe. This approach is also the basis of the GPS system.

The project for the current GPS system had been started by the *U.S. Department of Defense (DoD)* in the early 1970s, with the first satellites launched into orbit in 1978. Although the driver for GPS was its projected military use, there had always been a provision to also offer signal for civilian use, if that was deemed necessary later in the program.

One of the most ominous military reasons for being able to derive your exact position information anywhere in the world concerned the launching of *Polaris* nuclear missiles from submarines: if the launch location was not known precisely enough, it would be impossible to feed correct trajectory data to the missiles, and they could miss their intended targets by tens, if not hundreds of kilometers.

Nuclear missiles are not the kind of weapons that you would like to resort to having such a crap shoot with.

Another driver for GPS was to provide a superior replacement for the existing ground-based navigation system called *Long Range Navigation (LORAN)* that had been in use since the Second World War. The available location accuracy was expected to improve from about a couple of hundred-meter range with LORAN to just a couple of meters with the military-grade signal of the GPS.

The final incarnation of LORAN, *LORAN-C*, was intended to be used for military aviation and naval navigation, but with the advent of smaller receivers, enabled by the technology switch from vacuum tubes to transistors, LORAN-C was made publicly available and it was installed even in many commercial and private airplanes during 1970s.

The issue with all generations of LORAN was the fact that it used low frequencies for its ground-based transmitters, and these frequencies are highly sensitive to ionospheric conditions. Therefore, the accuracy of LORAN varied hugely depending on the weather and time of day.

Although President Reagan's order eventually did open GPS for civilian use, it still used two different modes: a high-precision signal for the military, and another, deliberately more imprecise, *selective availability (SA)* signal for civilians, which randomly reduced the accuracy so that it could be off up to 100 meters at times.

Despite this intentionally degraded signal, the use of GPS spread rapidly, as even the low-quality position information was far better than anything else at a comparable cost, and the required receivers were truly portable.

The final boost for GPS happened when President Bill Clinton unleashed its full potential with an executive order in 1996. He authorized the removal of the deliberate scrambling of the civilian signal. Therefore, in May 2000, the precision of all existing GPS receivers in the world improved more than tenfold without any extra cost or software updates.

This fundamental change happened at the same time as the advances in electronics made extremely cost-effective GPS receivers possible, leading to the explosion of navigation equipment use. The integration of GPS receivers has now reached a point at which even the cheapest smartphones offer in-built navigation functionality.

The GPS system is an intricate setup, consisting of 30 active satellites, plus several standby spares ready to step in in case of a failure, together with ground stations that keep the satellites constantly up to date and monitor their performance. At any given time, at least four satellites are visible at any location on Earth, and by listening to the encoded microwave transmissions from these satellites and referencing this information to the available data specifying their expected orbits, it is possible to calculate your position anywhere on Earth. The more satellites you can receive, the better the accuracy, down to just a couple of meters.

This seemingly simple arrangement is based on some extremely complex mathematical calculations, in which the three-dimensional "signal spheres" from the satellites are matched against each other to find your exact location, and it is the invisible magic resulting from this marriage of mathematics and radio waves that helps you to realize that you just missed a crucial turn on your way from A to B.

In a GPS system, the most expensive parts, including the very accurate atomic clocks that provide nanosecond (one billionth of a second) resolution timing information for calculations, are built into the satellites. Having the most complex parts of the system embedded in the supporting infrastructure makes it possible to have GPS receivers on a microchip that costs only a few dollars. As a result, precise location detection functionality can therefore be added cheaply to almost anything that moves.

Microchips are explained in TechTalk *Sparks and Waves*.

The GPS system initially became operational in 1978, and the first satellite of the current GPS generation was launched in 1989. With the addition of *Wide Area Augmentation System (WAAS)* satellites, GPS is now extremely precise. WAAS satellites send additional corrective information that helps counteract dynamically occurring signal reception errors, caused by random ionospheric disturbances.

There is an inherent limitation with basic GPS receivers when they have been powered off and moved to a new location: in order to realign itself, the receiver has to first scan the used frequencies in order to determine which satellites it is able to receive, and thereafter *download* the necessary satellite position tables, *ephemerides* and *GPS almanac*, to be able to perform any positional calculations. This data is repeatedly streamed down as part of the received signal, but as the data speed is only 50 bits per second, it may take up to 12.5 minutes until the receiver is again able to determine its location and provide a solid position fix.

TechTalk *Size Matters* discusses data speeds in detail.

This delay can be eliminated if there is a way to load the satellite location data tables faster into the GPS receiver, allowing the receiver to know almost instantly which satellites on which channels it is supposed to listen to. For this purpose, an extended GPS system, which is known as *Assisted GPS (A-GPS)*, is the norm in modern smartphones and is the reason why your phone can find your location the moment you turn it on, even after an intercontinental flight.

For aviation, GPS now offers another, inexpensive and extremely accurate source of navigation, especially when combined with a moving map technology. With a GPS moving map display, the kind of mistake that doomed *KAL 007* would have been obvious to detect, as the plane's deviation from the designed route would have been clearly visible.

The existing VOR, DME and ILS stations are relatively expensive to maintain, as they require constant monitoring and frequent testing for safety purposes. Hence the current trend is to move more and more towards *satellite-based navigation systems* in aviation.

The unprecedented precision of GPS with WAAS support has made it possible to create *virtual approach procedures*, closely resembling the ones provided by the ILS system. As these do not require any hardware at the designated airports and are therefore totally maintenance-free, they have created safer means for airplanes to land at even small airports in poor visibility. More importantly, in addition to making thousands of airports more available, these new, GPS-based approach procedures have made it possible to decommission many expensive-to-maintain ILS approach systems from less active airports, thus maintaining the same level of service with zero maintenance cost.

The same fate as with low-traffic ILS systems is expected for many of the VOR stations—only a bare-bones backup grid for the standard airways will be maintained in the future, as waypoints can just as well be defined via GPS, with much greater flexibility than by using only VOR radials.

The GPS system has made navigation cheap and easy worldwide, on land, at sea and in the air, and has created major cost savings and safety improvements. Additionally, having accurate location information universally available around the globe has created all kinds of new services and businesses that were impossible just a couple of decades ago.

But it has also injected a massive single point of potential failure in the system: as the GPS signals are received from satellites that orbit at an altitude of over 20,000 kilometers, the received signals are extremely weak and thus vulnerable to any interference on the same frequency band. This was an issue during the Persian Gulf War in 1991, when Iraqi forces used GPS jammers around sensitive military targets, and it is unlikely that GPS could be fully relied upon in any modern event of active warfare: even in our current civilian times, there has been reports of huge deviations being noticed in various parts of the Russian Federation, including near the Kremlin in Moscow, as well as on the shores of the Caspian Sea. Similarly, GPS reception at Newark airport was repetitively disrupted by a passing pick-up truck in 2013, as the driver was using a GPS jammer to hide his whereabouts from his employer. He was eventually caught and fined 32,000 dollars for his efforts.

In the unfortunate reality of constantly competing military forces, a system controlled by the armed forces of a single, potentially hostile country will naturally not be to the liking of everyone. As a result, an alternative system called *GLONASS* was created by the Soviet Union, and it continues to be maintained by Russia. China has its own *BeiDou* satellite system in orbit, originally targeted only to cover mainland China but with global expansion plans. Similarly, India has its own regional system, covering an area around the Indian Ocean from Mozambique to Northwestern Australia and all the way to the Russian border in the North, while Japan is working on their super-accurate, regionally optimized GPS enhancement, *Quasi-Zenith Satellite System (QZSS)*, expected to be fully operational in 2018.

Most recently, the *Galileo* system by the *European Union* and the *European Space Agency (ESA)* went live in 2016, after seventeen years of development worth 11 billion dollars, only to be hit by issues with the super-accurate atomic clocks onboard the satellites. It should be in full use in 2020, ultimately surpassing the precision level offered by the GPS system. Like GPS and GLONASS, Galileo is offering global coverage.

Most modern GPS receivers can synchronize with some functional combination of these parallel systems, hence providing redundancy for the users against a major malfunction or deliberate disconnect in one of them. Yet all of these systems are based on satellites, which are in danger of being hit simultaneously by a major *solar storm*, the worst of which in our recorded history, known as *The Carrington Event*, happened in 1859. There has not been anything comparable during our recent Space Age, and therefore we have no knowledge of how an event on an equal scale would affect not only GPS, but other satellites as well—all we know is that *something* will certainly break, but we have no idea of how widespread the devastation will be.

LORAN was deemed to become obsolete in the aftermath of making GPS available for all, and most European LORAN transmitters were shut down in 2015. Occasionally there has been some renewed interest for an upgraded *Enhanced LORAN (eLORAN)* version, which would function as a backup in case of catastrophic GPS outage.

The promised advantages of eLORAN are based mainly on improvements on the receiver technology, reaching an accuracy of tens of meters. Despite many suggestions around the world, at the time of writing this, the only countries still actively working on eLORAN are Russia and South Korea.

Although the use of moving maps on our smartphones today is amazingly easy from the end users' point of view, the technology behind generating that location information is by far the most mathematically complex use of radio waves ever designed by mankind.

So far.

Finally, the multitude of overlapping systems that are offering basically the same service is a sad reminder of how we are still bitterly divided into nation states, even though we all share the same *pale blue dot* in space. Running each of these services costs millions per day, and GPS alone has cost over 10 billion dollars to build so far. The estimate for the cost of running Galileo for 20 years has gone up from about 9 billion dollars to 25 billion dollars.

Even though the demand for high-precision location information is global and shared by all nations, there is no push to establish a shared system that would be managed by some neutral body like the *United Nations* (*UN*). As long as there is a clear military use for a service, honest collaboration between the artificial tribes of the world appears to be impossible.

In many other areas, though, international collaboration in the utilization of space has been extremely effective: one particular success story is the *Cospas-Sarsat* program, which has over 40 satellites in various orbits, listening to *406* MHz *distress beacons*. These *Emergency Position Indicating Radio Beacons* (*EPIB*) are standard equipment in ships and airplanes, but can also be purchased as portable *Personal Locator Beacon* (*PLB*) units for individual use.

*Cospas-Sarsat* satellites immediately detect any 406 MHz signal, emanating from anywhere in the world. The transmission contains digitally transmitted information consisting of the originator ID and the GPS-derived location. Most airplanes have EPIB-equipment on board, as a self-contained unit, connected to an internal battery and an impact sensor that activates the beacon automatically after an accident.

As the GPS position in the message might not be up to date, the satellites also perform their own location detection algorithm based on the *Doppler effect*, like the "reverse Sputnik" approach discussed above, and then relay all information to the several tens of automated ground control stations, which transfer the information to regional *Mission Control Centers* (*MCC*).

The system has been operational since 1982, and currently there are more than 550,000 registered EPIB and PAB devices. On average, more than 2,000 persons are rescued annually by the system, in over 700 emergency events around the world.

*Cospas-Sarsat* is a perfect example of functionality in which the user interface of a device is extremely simple: the triggering happens either automatically through sensors or by a button press. But everything that follows thereafter is very complicated: numerous satellites receive the signal, calculate positions based on complex mathematics and then relay all information to MCCs scattered around the globe.

This is an amazing example of how the harnessing of these invisible waves has made it possible to create a worldwide system that saves lives daily.

# Chapter 7
# Traffic Jam Over the Equator

Getting into space is tough.

Gravity tries its best to ensure that what goes up, must eventually come down: the only way to stay up is to move fast enough forward so that the Earth's curvature bends away under you while gravity relentlessly keeps on pulling you towards the Earth. You have to literally keep falling off a cliff, except your cliff is the horizon that you will never reach, as it is curving down in sync with you.

Achieving this endless fall over the horizon requires that the satellite must be accelerated to a speed of at least 8 kilometers per second.

Not per hour or even per minute. *Per second.*

That's like covering the distance between London Heathrow and Paris Charles de Gaulle airports in less than 45 seconds, or crossing the Atlantic from London to New York in just under 12 minutes.

This kind of orbital speed requires vast amounts of energy, and therefore the actual payload of modern rockets is just a tiny portion of the total structure that starts lifting up at takeoff: most of the size and weight is needed for the fuel that accelerates the relatively small mass of the satellite at the top of the rocket. Despite the recent amazing advances in reusable rockets by companies like *SpaceX* and *Blue Origin*, going into space is still a very costly business: the launch costs, even without counting the value of the actual satellite or satellites on board, are well into several tens of millions of dollars.

But the multitude of advantages with having satellites constantly over our heads make all this effort worthwhile, and hence launching satellites has been a booming business over the last fifty years.

There are more than 4,000 active satellites orbiting the Earth at the time of writing this, and about the same number of satellites have ceased to work, either creating a continuously growing set of orbital *space junk*, or being destroyed as spectacular streaks of fire in the sky after re-entering the atmosphere.

© Springer International Publishing AG, part of Springer Nature 2018
P. Launiainen, *A Brief History of Everything Wireless*,
https://doi.org/10.1007/978-3-319-78910-1_7

Anything that remains uncontrollable on orbit needs to be constantly monitored for potential collisions: for example, the only currently manned space station, the *International Space Station* (*ISS*) has to make frequent minuscule adjustments to its orbit in order to avoid hitting these rogue objects that cross its path.

And the number of orbiting satellites is constantly increasing—just one launch by India in early 2017 carried over 100 tiny *nanosatellites* into orbit.

In terms of being able to communicate with the satellites, the huge relative speed between a satellite and any ground station causes certain complications:

Satellites on a *Low Earth Orbit* (*LEO*), which refers to altitudes below 2,000 kilometers, fly around the Earth in less than two hours, repeatedly passing from horizon to horizon over the locations below their orbits. Therefore, having a constant radio connection between a ground station and a satellite requires multiple ground stations scattered around the Earth, so that at least one of them has the satellite in sight. And as the transmission power of a satellite is often very low due to the limited amount of power provided by the solar panels, it is usually necessary to utilize a highly directional, movable antenna that can track the position of the satellite in real-time.

In addition to this, the constantly varying relative speed between a ground station and a satellite in LEO also causes the communication frequencies to shift according to the *Doppler effect*, which has to be continuously corrected for. The positive uses of this effect were discussed in Chapter 6: **Highways in the Sky**.

All in all, these inherent properties make the process of tracking LEO-based satellites quite complicated.

The very first communications satellite, *Telstar*, became operational in 1962 and enabled intercontinental live television transmissions between Europe and North America, but due to its low-level orbit around the Earth, the intercontinental communications window was only available for 20 minutes at a time, and occurred only once in every 2.5 hours.

Despite this serious limitation, *Telstar* heralded the start of a new Space Age of communications, connecting audiences across continents in real-time. To commemorate this achievement, an all-instrumental pop song carrying the same name was composed by a band *The Tornados*, becoming the first British single that reached the number one position in the USA.

The Space Age had a mind meld with Pop Culture.

The narrow intercontinental transmit window was not the only limitation of *Telstar*: the available transmit power was also so small that the tracking antenna in Andover, Maine had to be enormous, about the size of a bus, and weighing over 300,000 kg. As the position of the satellite changed with a speed of 1.5 degrees per second, the moving antenna platform that was needed for tracking *Telstar* was a mechanical miracle in itself.

There is, however, one orbit that is very special, completely removing the requirement of continuous tracking and the issues of *Doppler effect* in frequency: if you position the satellite at an altitude of 35,786 kilometers and align the orbit with the equatorial plane of the Earth, the time it takes for the satellite to do one rotation around the Earth will be exactly 24 hours.

Therefore, the orbital speed of the satellite matches the speed at which the Earth is rotating around its axis, and the satellite seems to remain continuously in the same position relative to the horizon. If you point your antenna towards it once, you do not need to adjust your antenna ever again: whether you are transmitting or receiving, your connection with the satellite remains available from the same fixed location in the sky. And as the relative speed between your antenna and the satellite is zero, there is also no *Doppler effect* to worry about.

An extra bonus also results in from such a high orbit: any satellite at this altitude remains visible from about a third of the Earth's surface at a time. Hence, by putting a transmitter on the satellite, you can theoretically cover all receivers residing anywhere over about a third of the globe in one go. The further North or South you move on the Earth, the lower in the horizon these satellites appear to be, and thus only areas close to the North and South Poles remain outside of the theoretical coverage, most often due to masking high terrain on the horizon.

The application of this kind of *geostationary satellites* completely removes the need for complex, moving antennas that would otherwise be needed to maintain constant communications if the satellites were to reside on a Low Earth Orbit—you only need to align your antenna once and then communicate happily ever after.

The mathematics behind the rotational speeds of various orbits were understood at the beginning of the 20th century, and the existence of such a *geostationary orbit* at roughly 36,000 kilometers altitude had been first noted by Russian scientist Konstantin Tsiolkovsky, who was one of the pioneers in the theory of rocketry. The first reference to the potential of this kind of orbit for communications purposes can be found in a book by another early pioneer of rocketry, Slovene Herman Potočnik, published in 1928, just one year before his unfortunate early death at the age of 36.

The very first in-depth discussion on the benefits of a geostationary orbit for wireless relays and broadcasting purposes can be traced back to an article entitled *Extra-Terrestrial Relays—Can Rocket Stations Give Worldwide Radio Coverage*, published in the *Wireless World* magazine in 1945. It was written by a young British engineer called Arthur C. Clarke, who later became a very prolific *Science Fiction* writer, correctly predicting many future developments, as will be discussed in Chapter 8: *The Hockey Stick Years*.

Clarke's article came out in the post-Second World War era, at a time when there was already a glimmer of hope of achieving real space flight, thanks to the war-time experiences with the German *V-2* ballistic missiles. But as the state-of-the-art in wireless communications was still based on vacuum tube technology, Clarke's solution expected that maintaining such relaying space stations in working condition would require constant human presence in orbit.

It took almost twenty years until the first practical application of the geostationary orbit was finally in place: the *Syncom 3* satellite proved the usefulness of such a setup by relaying live television broadcasts from the 1964 *Summer Olympics* in Tokyo to viewers in the United States. Unlike with *Telstar*, the intercontinental connectivity provided by *Syncom 3* was continuous, and there was no need for massive, constantly moving antennas.

A rush to exploit this unique orbit then followed, and the geostationary *Clarke Orbit*, as it is sometimes called, is now getting very crowded: currently there are over 400 satellites circling in sync with Earth's rotation over the Equator.

All known satellites around the Earth are beautifully visualized at an interactive website:

*http://bhoew.com/sat*

The special ring of geostationary satellites stands clearly out from the crowd.

Due to the potential interference between adjacent satellites, their positions, or *slots* as they are called, and the frequencies used for communications, are strictly regulated by the *International Telecommunication Union* (*ITU*).

Many of the geostationary satellites are used for direct broadcast satellite television, but some act as communications links between continents or offer rentable channels for whatever purpose the customer wants to use them.

As an example, most mobile camera crews of live television operations use satellite connectivity to send their field transmissions back to the company headquarters, and most live intercontinental news reports go across at least one satellite on the Clarke Orbit.

Other notable uses for these geostationary slots are weather satellites that can continuously keep track of a third of the Earth's surface at once, the satellite radio network *Sirius XM* that provides uniform digital audio and specialty data reception of hundreds of channels across the continental United States, *Wide Area Augmentation System* (*WAAS*) support for enhanced GPS precision, as explained in Chapter 6: *Highways in the Sky*, and inevitably, also several satellites that are used for military communications and surveillance.

The *International Space Station* (*ISS*) is kept on an orbit that varies in altitude between 330 and 425 kilometers, and hence resides on a Low Earth Orbit, but the communications facilities for *ISS* rely on *NASA's Tracking and Data Relay Satellite* (*TDRS*) system. It is based on satellites sprinkled over the Equator, so that as the *ISS* whizzes around the world in about 90 minutes, there's always at least one *TDRS* satellite visible and thus available, providing uninterrupted two-way communications with the *ISS*.

Thanks to this high-bandwidth connectivity, there are now *YouTube* channels that provide constant video feed from the activities and views from the *ISS*, easily found with the search term "NASA live stream".

As mentioned above, there are several weather satellites on the geostationary orbit, and a nice example of the excellent quality of the weather satellite coverage that can be achieved from that kind of vantage point, in real time and recorded in various wavelengths, can be found here:

*http://bhoew.com/wea*

This website updates every ten minutes, providing the newest picture taken by the Japanese *Himawari* satellite, which covers the globe from Kamchatka Peninsula in the North to Tasmania in the South.

Satellites use *microwave* frequencies for communications, which, due to their high frequency, accommodate a multitude of individual channels into a single transmission beam, for reasons explained in TechTalk *There is No Free Lunch*.

The flip side of this high frequency is the fact that water absorbs microwaves very effectively, so hard rain or intense snowfall can therefore cause problems in reception.

Further discussion about microwaves can be found in Chapter 11: **Home Sweet Home**.

Another less obvious issue with geostationary satellites is the fact that any communication with them must be able to traverse a distance of 36,000 kilometers: this is almost one tenth of the distance to the Moon and just 4,000 kilometers short of the circumference of the Earth, and due to the laws of physics, this causes some limitations.

First of all, as the power of your transmitted signal drops to one fourth of its original strength every time you double your distance, you need a lot of transmission power to produce a signal strong enough to be received at the other end. This can be remedied to some extent with highly directional antennas, which are the norm for satellite communications—everyone is familiar with the multitude of television dish antennas scattered around our cities and towns.

Even though it is theoretically possible to cover about a third of the Earth's surface with a single satellite, having a satellite television transmitter powerful enough and with a wide enough beam to achieve this is currently not a viable solution, unless the receiving antennas are enormous. Due to this, the direct broadcast television satellites have highly directional antennas that cover only specific areas of the Earth: typical beams can cover an area the size of Central Europe, for example, and the further away you get from this optimal reception area, the larger your antenna has to be to counteract the reduced strength of the signal.

The second limitation is much less obvious:

Because radio signals propagate with the speed of light (about 300,000 kilometers/s), a distance of 36,000 kilometers is long enough to create a noticeable delay. This delay becomes even more pronounced if you first send the signal out and it then gets beamed back, making a total roundtrip of 72,000 kilometers.

I remember experiencing my first concrete example of the physical limits of the speed of light when I was a kid, watching a documentary that was describing the Canadian *Inuit Broadcasting Corporation (IBC)* satellite television system, which provides television transmissions to the Arctic areas of Canada. The documentary had a section in which there were two television monitors side by side, one showing the uplink transmission that was being beamed to the satellite, and the second one showing the same program as it was received back from the satellite.

Because the total distance from first sending the signal to the satellite and then retransmitting it back was two times 36,000 kilometers, this caused a clearly noticeable delay of 0.24 seconds between the monitors.

This very expensive "delay line" is a good example of how it is one thing to read about theoretical scientific subjects like the speed of light, and very different to see what it means in practice.

Seeing is believing.

For unidirectional broadcast purposes, like transmitting a live sports event across continents, a delay of 0.24 seconds is nothing, and most modern digital television

systems easily add more to that due to the extra time it takes to buffer and decode
the received signal, but there are many interactive applications where a delay like
this becomes noticeable:

Next time you watch a live newscast, note the delay between the question from
the studio and the corresponding answer from the field.

The cause of this unnaturally long pause is not due to the reporter having to think
hard before responding—it is simply due to the time it takes for the question made
in the studio to make its way up to the satellite, bounce off of it to the field crew
receiver, and then the same amount of up-down trek again as the field reporter starts
to respond to the question.

All in all, this amounts to about half a second, not counting the additional delays
in processing the actual digital signal.

If you have a live connection between, say, the United States and Australia, you
may end up having multiple hops across several satellites, adding even more delay
to the connection.

In any case, you are witnessing a concrete example of the most fundamental
speed limit of the universe in action, and for any communications that require
interactivity, like simple intercontinental phone calls, this extra delay that is caused
by the 72,000 extra kilometers' roundtrip soon becomes very annoying.

To counteract this, the world's oceans are being crossed over and over again by a
multitude of *fiber-optic data cables*, which, thanks to their much shorter
point-to-point connections, reduce the speed of light delay to levels that have no
noticeable side effects in most real-word use cases.

As mentioned, there are only a limited number of interference-free slots available on this geostationary orbit, and due to *solar wind* and minute variations in
gravitational forces, some even caused by the masses of neighboring satellites,
geostationary satellites need to make repetitive positional corrections in order to
remain in the exact location. Eventually there is no longer enough propellant left for
these corrections, and the satellite is usually decommissioned by pushing it about
300 kilometers further out to the so-called *graveyard orbit*, so that it won't end up
on a collision course with its still-functioning neighbors.

This end-of-life procedure, requiring about the same amount of fuel as three months
of normal position keeping on the Clarke Orbit, has been a mandatory requirement for
geostationary satellites since 2002, and it guarantees that the satellite can eventually be
moved to a location in space in which it is not interfering with other satellites. Despite
this regulation and earlier, voluntary approaches to identical end-of-life management of
satellites, there are over 700 uncontrolled objects now adrift close to the geostationary
orbit, either due to a system failure or by lacking an end-of-life procedure, and they all
need to be constantly tracked to avoid collisions.

In most cases, after decommissioning or losing an existing satellite, the freed slot
will be taken over by a brand new and much more capable satellite with shiny-new
technology, often already planned for this replacement task several years in
advance, as these locations are limited and hence very valuable. Handling the
allocations of these slots is one good example of successful international collaboration regarding shared "real estate" in space.

Despite its numerous advantages, the geostationary orbit is by no means the only one that is in use for satellite-based communications: if you need to have a handheld system on the ground that needs to be able to communicate with a satellite, the 36,000 kilometers distance requires way too much transmission power, and the delay caused by crossing this huge distance twice may be unacceptable for interactive uses. Therefore, if you want to combine true portability with true global coverage, you need to create a different solution.

Meet *Iridium*—an ill-fated communications phoenix that was way ahead of its time at its launch in 1998, and rose back from the ashes after one of the costliest bankruptcies in history. *Iridium* has satellites orbiting at about 800 kilometers altitude, and it got its name from the fact that the original planned number of required satellites was 77, which is the *atomic number* of Iridium, a silvery-white metal. Even though the number of satellites was eventually reduced before the launch, the catchy name stuck.

*Iridium* was the brainchild of the notorious *Motorola Corporation* and ended up costing about six billion dollars before it went bankrupt in almost immediately after its commercial launch in 1999. It first continued operating under the American *Chapter 11* bankruptcy protection, until in 2001, a set of private investors bought the rights to the existing system and the orbiting satellites for only 35 million dollars.

This was a very decent 99.4% discount for a fully functional worldwide communications system.

After a couple of mergers to strengthen their financial position, the new owners revitalized the system and are now continuing to offer a satellite-based, portable phone and data service that seamlessly covers the Earth.

Because the original satellite technology is from the 1990s, the supported data speeds are very slow by today's standards, about 2.4 kbps, with a time-based cost of roughly a dollar per minute at the time of the writing, both for data and voice connections, so using *Iridium* is by no means a cheap affair, but if you need true worldwide coverage, this solution works from pole to pole:

An example of a connectivity solution via an *Iridium*-based data link is the setup that has been used to connect the *Amundsen–Scott South Pole Station* to the rest of the world, with parallel *Iridium* modems providing a 28.8 kbps speed—just about on par with the early *acoustic modems*.

Due to its extreme location at the South Pole, the station is not able to have a line-of-sight connection with standard geostationary satellites: it relies on a couple of special satellites which keep on crossing over and under the equatorial plane while still maintaining a 24-hour *geosynchronous* orbit. This provides a high-speed satellite communications window during those periods when the connecting satellite moves to the Southern side of the Equator, hence becoming just barely visible over the horizon as seen from the South Pole.

Most of this kind of high-speed connectivity provided for the Amundsen-Scott South Pole Station comes from a selection of *NASA's Tracking and Data Relay Satellites* (*TDRS*), but as they keep on "wobbling" over and under the equatorial plane, this high-speed connectivity is not available all the time.

Therefore, the *Iridium*-based setup is the only backup link that offers continuous availability, and thanks to recent advancements, the researchers in Antarctica will soon be able to have an upgrade: with its initial stellar debts having been washed away through the bankruptcy fire sale, the reborn *Iridium* company has been profitable enough to start renewing their satellites with the latest communications technologies, and as a result, is able to offer considerably improved communications capabilities.

The maximum data speed for the new *Iridium Next* system and its totally revamped *Iridium OpenPort* data configuration will increase to 512 kbps. Existing *Iridium* handsets and modems will continue to work with the new satellites, but with the state-of-the-art satellite technology and matching new devices, *Iridium Next* will be able to offer not only higher data speeds but also many new features, like 64 kbps broadcast mode for multiple devices, together with live tracking support of ships and airplanes.

The first user of the new real-time airplane tracking feature offered by *Iridium* is *Malaysian Airlines*, which is still reeling from the unexplained disappearance of their flight *MH 370* in 2014.

The first batch of these new *Iridium Next* satellites was launched in January 2017, on a much-anticipated launch of the *Falcon 9* rocket by *SpaceX* after the spectacular explosion that destroyed the previous one in late 2016.

This time, though, everything went by the book, including the soft landing and recovery of the *Falcon 9* first stage, and most importantly, having ten new *Iridium Next*-satellites on their expected orbits.

*Iridium* has also achieved another, not-so-positive history first: in February 2001, one of the *Iridium* satellites collided with a defunct Russian *Kosmos 2251* satellite. The relative speed of these two satellites at the time of impact was estimated to be roughly 35,000 kilometers/h, and the collision created an enormous and potentially destructive debris field in orbit. This event shows the importance of tracking all objects in space: the worst-case result from a set of collisions is a chain reaction that could destroy tens of other satellites and make huge parts of potential orbits unusable for decades.

The original *Iridium* system has one interesting and surprising unplanned side effect: because the microwave antennas on the satellites are large, polished slabs of aluminum, in the right conditions they act as mirrors, reflecting the light of the Sun to locations on Earth which are already in the deep darkness of the night. Therefore, for anyone at the right place at the right time under the clear night sky, this creates an experience of a slowly moving star suddenly becoming visible, bursting into a star brighter than any other star in the sky for a short while, and then vanishing again.

The occurrences of these *Iridium flares* can be calculated for any location on Earth, and being able to "predict" where and when a "new star" appears can be used as a great source of amazement if you have kids: the website for calculating the timing of Iridium flares for any location on Earth can be found at:

*http://bhoew.com/iss*

Give it a go, but hurry up—the new *Iridium Next* constellation will no longer have similar antennas, so this free, heavenly spectacular will gradually go away around 2018–2019.

The same website contains tracking information of several other visible satellites as well, including the *ISS*, which is by far the largest object in orbit and hence another impressive sight in the night sky. Its orbit stays fairly close to the Equator, though, and hence it is not visible on high Northern and Southern latitudes.

Another space communications pioneer, *Inmarsat*, has been offering satellite-based voice and data connections since late 1970s. Created originally as a non-profit international organization for maritime use, *Inmarsat's* operational business became a private company in 1999, and has changed ownership several times since then.

The *Inmarsat* solution is based on satellites on geostationary orbits, and the resulting delay means that they are best suited for cases where the delay has no adverse effect, like for email, browsing or unidirectional content streaming. *Inmarsat's* latest high-speed service, *Global Xpress*, has one additional special feature: it offers steerable antennas, allowing highly localized, high-capacity services to be offered on demand on a global scale.

Several companies have been offering satellite-based Internet for decades, including *Inmarsat*, *HughesNet*, *ViaSat* and *europasat*. Some of these companies are also behind the latest fad in intercontinental travel: in-flight Wi-Fi connectivity, as satellites are the only option for seamless communications over the oceans.

Many of these satellite providers are the only Internet connectivity option for rural customers, and usually their plans have some serious limitations in terms of the amount of data downloads that are included in their monthly charges. But as the technology has improved, there is now renewed interest in utilizing the Low Earth Orbit for communications purposes.

Despite its recent upgrades, the LEO pioneer *Iridium's* newest *Iridium Next* incarnation will pale in comparison with the plans that the most versatile innovator of recent history, Elon Musk, has for the next couple of years: his company, *SpaceX*, in not only revolutionizing the launch business with his reusable rockets, but also it plans to implement a LEO-satellite Internet service, *Starlink*, which in its first stage uses no less than 4,425 satellites to cover the globe, expanding up to 12,000 in its final configuration. The tests are due to start in 2018, and the full deployment should be ready by 2024, with the aim of first offering fast Internet access to the United States, and expanding to a global service in later years.

The system is planned to provide gigabit speeds, which would compete with *fiber-optic data cable*-based terrestrial Internet providers, and in order to further improve the throughput of the new system, the satellites also implement a "celestial" *mesh network*, the concept of which is further discussed in TechTalk **Making a Mesh**.

The plan is nothing short of ambitious, as the number of new satellites would quadruple the current number of existing, active satellites in space, but with his proven technological revolution in electric cars and solar energy systems by *Tesla* and reusable rockets by *SpaceX* in the satellite launching business, Elon Musk's track record gives a lot of credibility to this endeavor. Having giant companies like *Google* and *Fidelity* financially backing his attempt by billions of dollars will also help.

Many traditional aerospace companies, like *Boeing* and *Airbus*, have also presented their satellite-based Internet plans, both privately and in connection with existing players. The resulting competition between all these newcomers and the incumbent players can only mean one thing for us consumers: more products to choose from and more potential for truly universal connectivity, with higher connection speeds and lower prices.

And finally, in the current atmosphere of ever-increasing attempts to discredit scientific approach by so many prominent political leaders, it gives me renewed faith in humanity when billionaires like Elon Musk and Bill Gates put their wealth to work to improve our lives and advance the global reach of technology and health care. This kind of behavior is a fresh, welcome contrast to that of so many recent *nouveau riche* who use their newly-acquired wealth just to buy sports teams with highly inflated price tags.

# Chapter 8
# The Hockey Stick Years

In 1987, the first major cracks in the fabric of the Soviet Union were starting to become visible, and Mikhail Gorbachev, the Soviet leader, had been painfully aware of this for quite some time.

Crude Oil exports were the main source of foreign currency for the Soviet Union, but the price of oil had collapsed from nearly 120 dollars per barrel just a couple of years back, to barely over 30 dollars per barrel now.

The income of the Soviet Union was therefore squashed but the costs were as high as ever: Cold War military spending was way over the level that the ever-weakening economy could support, and the aging domestic industry was in dire need of crucial and costly structural improvements.

The current situation in Russia at the time of writing this book is uncannily similar to what the Soviet Union was experiencing in 1987, except that in 1987 Ronald Reagan was in the White House, and he did not feel any sympathy for the growing plight of the Soviet leadership at the time.

In order to avoid running out of money, the Soviet Union had to try to attract foreign investments and open new avenues of technological collaboration with foreign companies. To accomplish this, Gorbachev embarked on a new, softer political direction, *Glasnost*, which aimed at opening Soviet businesses to the West and reducing the costly political and military tensions, the origins of which could be traced all the way back to the end of the Second World War.

Undoubtedly, Gorbachev had been watching the economic progress in China with strong feelings of envy and admiration: Deng Xiaoping had spent years repairing the damage left behind by Mao Zedong, and under his reformist lead, China had transformed itself into a country with impressive economic growth while still managing to keep the strong hand of the Communist Party at the helm. The situation in China was kind of a Soviet dream but in a country that only a few decades ago had been devastated by domestic political mismanagement and incompetence of an almost unimaginable scale. But thanks to Deng Xiaoping's reforms, China had made an impressive turn from an underdeveloped and rural communist backwater to a budding industrial powerhouse, squarely aiming at the world market.

© Springer International Publishing AG, part of Springer Nature 2018
P. Launiainen, *A Brief History of Everything Wireless*,
https://doi.org/10.1007/978-3-319-78910-1_8

Gorbachev, with his back against the crumbling wall of the Soviet economy, was determined to push for an identical change in the hope of creating enough positive economic inertia that could help to bridge the gap caused by the crash of the price of oil. An important part of his new drive with Glasnost was to improve relationships with the closest neighboring countries, and due to this, he visited Finland in October 1987.

The pretext of this visit was the signing of a declaration that unequivocally categorized Finland, a country that the Soviet Union had twice tried and failed to take over during the Second World War, as a neutral country. Along with this political issue, an important part of the program was to come to an understanding about potential business collaboration opportunities with Finnish companies and academia.

Little did he know that there was a plot to make him a hapless, free advertiser of an obscure Finnish company—a company that had recently made a groundbreaking transformation from a sprawling, tired conglomerate into an ambitious telecoms-focused company.

This company was *Nokia*.

During a press conference right after a presentation on Finnish industry and technological research facilities, Gorbachev was casually handed a gray piece of equipment with a distinctive antenna on top and told that "there's a line open to Moscow" to his Minister of Communications. Perplexed and totally taken by surprise, he started having a conversation, while the accompanying journalists photographed and filmed him.

As a result, the gray device in question, the official name of which was *Mobira Cityman 900*, came to be known as *Gorba* from there on.

In terms of its external design, *Gorba* looked quite similar to the world's first portable mobile phone, *Motorola's DynaTAC 8000X*, which had been launched in 1984, just three years earlier. But what made *Gorba* tick was the wireless connectivity provided by the *Nordic Mobile Telephone (NMT)* system, which was the first fully automatic *cellular phone* system available across national borders. Initially at its launch in 1981, it covered Sweden and Norway and was expanded to Finland and Denmark the following year.

The network was originally designed for in-car use, but with the advances of technology, devices like *Cityman 900*, having a hefty weight of almost a kilogram, became available in a portable enough form for personal use.

Although NMT was a brand-new network, it was a result of a long line of earlier research into cellular technology: the joint group of Nordic telecoms was set up in 1969, and it followed the path laid down by the developments that originated in the United States, where the legendary *Bell Labs* had introduced the concept and had also performed the first large-scale practical trial.

The concept of *cellular network* offered a novel approach to overcome some of the restrictions of traditional radio communications, the most fundamental of which were the limited number of channels available, and the inevitable battle between portability and achievable range for two-way communications. More power means longer range, but requires larger batteries in the portable handsets, which soon become prohibitively heavy and expensive, thanks to the laws of physics in the form of the *inverse square law* that affects the propagation of radio waves. This means that in order to double your range, you have to quadruple your transmission power. Finally, with longer range you hit another undesirable side effect: the particular transmission channel in use cannot be reused again within the whole coverage area, as any other transmission on the same channel would cause interference.

This leads to the inevitable battle between the desired range and the number of simultaneous communications possible within the limited number of available channels. If you attempt to cover an area like an entire city with your new wireless service and your service becomes successful, you soon use up all the channels available for simultaneous calls, and adding more users would seriously degrade the performance of the system.

The earliest mobile networks had to live with these grave limitations: for example, the *Improved Mobile Telephone Service (IMTS)* network of *AT&T* in 1965 had only 12 channels available for 2,000 users in the New York area, and the average wait time for making a call that was manually connected by a human operator was 20 min.

To work around this apparently insurmountable problem of limited channels, a proposal had been made in 1947 by Douglas Ring at *Bell Labs*. He suggested that instead of using powerful central transmitters to provide city-wide coverage, a set of low-power *base stations*, divided geographically into small *cells,* would be used to provide highly localized connectivity instead. These base stations would only provide service to those customers that happened to reside within the limited coverage area of the cell.

The benefit of this would be that even though these base stations would all use the same frequency band, it was possible to distribute the actual allocated channels intelligently between the base stations in the whole coverage area in such a way that no adjacent base stations shared the same channels. Due to the lower power needed to only cover one cell area, more distant base stations could therefore reuse the same channels again, without having to worry about possible interference resulting from simultaneous use of the same channel somewhere further away.

As a consequence of this approach, the limitation for the number of users that can be served at any one time becomes *local to the cell* instead of being *global to the whole coverage area*, and thanks to the shorter distance coverage required from the connecting devices, the maximum required transmission power for handsets could also be reduced, which enabled the use of smaller handsets with cheaper and especially lighter batteries.

Although this cellular approach neatly solved the most crucial problem of maximum simultaneous use of a limited set of channels and therefore enabled even a nationwide coverage area by constantly reusing the channels, it created a new, even more complex problem: as the connected user was moving away from its currently active base station, the handset had to be constantly aware of other available cells in the vicinity and be able to request the current base station to switch the ongoing call to another cell when that cell was determined to offer a better connection quality.

During this process, called *handoff*, the ongoing call has to be seamlessly switched to a new radio channel that connects the handset to the base station of the new cell, and the audio of the ongoing call needs to be rerouted from the previous base station to the new base station. And in order to avoid any noticeable disruption in the ongoing call, all this needs to happen in a time frame of 0.1–0.2 seconds.

Even though the proposed theory behind cellular networks was solid, this kind of on-demand routing of audio between base stations and the required dynamic radio channel switching in the handsets simply could not be resolved with the existing 1950s technology.

But as with the *super-heterodyne theory* in the early years of the 20th century, advances in electronics eventually made fully automated handoffs possible, thanks to the use of computers that could handle the necessary switching logic both in base stations and handsets.

The first large-scale implementation of a cellular network was set up by the very same *Bell Labs* in Chicago in 1978, over three decades after Douglas Ring had made his proposal, followed a year later by the *Nippon Telephone and Telegraph (NTT)* in Japan.

And barely two years later came the *Nordic Mobile Telephone (NMT)* system, providing the first network that worked transparently in multiple countries.

The required transmission power is not the only factor affecting the potential size of a device planned for a mobile network: first of all, the optimal antenna length is inversely proportional to the frequency in use, and therefore the higher the frequency, the more compact the antenna may be. Secondly, the necessary modulation needed to carry the transmitted signal defines the bandwidth needed for any individual channel, which in turn fixes the total number of channels that can be made available within the allocated block of frequencies for this purpose.

The fundamental limitation that stems from the need of bandwidth is discussed in detail in TechTalk *There is No Free Lunch*.

All these restrictions dictate that the useful frequencies for truly portable devices with enough channels to support hundreds or thousands of simultaneous users start from about 400 MHz, and as an example, the NMT system was originally set to operate on a 450 MHz frequency band, expanding later to also cover 900 MHz.

The fact that 900 MHz is double 450 MHz is no coincidence—optimal multi-frequency antenna design is easiest to achieve when the frequency bands in use are integral multiples of each other.

Because the available radio frequency spectrum is a limited resource, its use is highly regulated on both national and international level. Frequency allocations have been jointly agreed upon over the history of radio, and new inventions can't just jump in and start using frequencies that are already allocated for other uses. Instead, a new block must be reserved or a previous one reallocated for the new service, and standards that define the ways in which this block is going to be used must be agreed upon. Therefore, in order to enable new wireless services, inventing novel ways to do useful things is not enough—an enormous amount of negotiations and international collaboration is required before a new concept can be turned into a real, world-wide service.

Although the *microchip* revolution made these *first-generation (1G)* cellular phones possible, the available *microprocessors* were rather power hungry and the existing battery technology was far from optimal. For example, the active talk time of *Gorba* on full charge was only 50 minutes. But for many businesses relying on connectivity while on the road, this was more than enough—being able to make and receive calls from practically anywhere was well worth both the constant recharging and the heavy weight of the handset.

Microchips are explained in TechTalk **Sparks and Waves**.

To add to its perceived usefulness, the inbuilt cross-border design of NMT brought a totally new dimension into the game: at the time of its launch, it was truly revolutionary to be able to turn on your handset in a foreign country and still make and receive calls via your home country's phone number. In many ways, NMT was the test bed for several features that we now take for granted in our current mobile networks.

After the resounding success of these pioneering networks, similar systems started sprouting all over the world. The United States got their *Advanced Mobile Phone System (AMPS)* two years after the launch of NMT, while Europe went on to have nine, mutually incompatible systems during the late 1980s and early 1990s.

All these first-generation cellular systems were based on analog technology. As with any traditional radio, you were the sole user of the channel you were talking on, and with a simple *scanner*, a portable receiver that could tune into the channels used by the cellular network, your discussion could be eavesdropped by anyone.

And eavesdropped they were.

Many embarrassing scandals ended up being leaked to the press, simply by listening to the ongoing phone conversations at strategic locations. This got so bad that scanners that could be tuned into the analog mobile phone bands were made illegal in the United States.

Naturally, making something illegal in one country did not make it unavailable, so those who benefited from eavesdropping found the necessary means to continue doing it.

Having analog channels was not only insecure but also wasted the precious frequency spectrum. Unlike traditional radio conversations, where only one person could transmit at any one time, your cellular network connection was *full duplex*, meaning that you had to have two channels allocated for you for the entire duration of your call—one for the *uplink* audio that transmitted your speech to your counterparty, and another for the *downlink* audio that transmitted the voice of your counterparty to you.

To make it even worse, the channels needed to be relatively wide to accommodate the analog speech signal. Similarly, as with any analog radio signal, the quality of the call was linearly dependent on the distance to the base station, and was severely affected by any stray interference on the channel.

In terms of cost, using any of these first-generation mobile phones was still a luxury. *Cityman 900* had a list price of about 8,000 dollars in current value, and the actual airtime was almost prohibitively expensive, too.

Despite all these shortcomings, first-generation networks rapidly spread around the world, the cost of handsets and airtime started the inevitable downwards curve of all new technology applications that succeed in creating a true mass-market, and being able to make calls while on the move turned out to be the biggest success story of the 1990s.

But with success came the inevitable capacity problems.

The customer base quickly grew to a point in which some cells became heavily congested. Customers would experience increasing amounts of dropped calls or busy signals when trying to initiate a new call. Despite the cell-by-cell reuse, the relatively small total number of available channels became a severely limiting factor in densely populated areas.

The networks had again reached the point in which adding more users would only degrade the perceived user experience.

Salvation came in the form of ever faster and cheaper microprocessors and especially *Digital Signal Processors (DSPs)*, that could move the system from analog to fully digital domain. DSPs brought the added benefit of real-time *digital signal compression,* provided by specific *audio codecs*, which are smart programs written for DSPs.

These concepts are discussed further in TechTalk *Size Matters*.

The *second-generation (2G)* all-digital systems were still using the same frequency bands as the first-generation networks, so the available number of channels was not considerably altered. But instead of hogging the whole bandwidth of a channel for the full duration of the call, digital signal compression reduced the bandwidth needed to transmit audio for a single user, and as a result, the same physical channel could now be shared in real-time between several users by dividing it into short *time slots*.

This kind of *Time Division Multiple Access (TDMA)* technology resulted in a highly pulsed transmission, in which the repetitive high-frequency pulses created a low-frequency buzz that was well within the audible range. A side effect of this was that almost every audio device was susceptible to this kind of interference, which was the reason why you could hear a constant buzz when you placed your active 2G phone near to, say, an FM radio.

As the phone starts its handshake process with the base station a couple of seconds before the actual call is connected, your nearby audio equipment could therefore give you an audible "warning" of the incoming call moments before the phone actually started ringing.

The pulsed transmission mode and its potential for resulting audible interference was a real problem for designers of these new handsets, as the pulsating radio transmission also had a tendency to bleed over to the audio circuitry inside the phone. This was most prominent when using a wired headset connected to the phone. To minimize this, all mobile phones in the early days had proprietary headset connectors, forcing the use of a particular headset that was provided by the manufacturer. This combination was then optimized to reduce the audio interference caused by the high-power pulse-generating transmitter inside the phone itself.

Connecting external wires to a device that included a radio transmitter could also potentially affect the performance of the actual radio circuitry inside the handset, and hence the engineers trying to cope with these sporadic problems were demanding that only proprietary, purpose-built and electrically optimally coupled headsets should be allowed to be used.

But as cellular phones gained more inbuilt features and also became standalone music and video players, users wanted to be able to use their own, preferred headphones.

As usual, ultimately the customer dictates the desired features, and today nobody gives a second thought as to whether the handset and headsets are from the same manufacturer. The engineers eventually found a way to make the internal circuitry of a handset immune to these kind of problems, and despite the current push by *Apple* and *Google* to get rid of the headphone jack entirely, the venerable 3.5-millimeter audio plug keeps hanging on.

As discussed earlier, the most impressive new feature provided by the NMT system was automatic international roaming, although it only worked between the four Nordic countries. This feature became one of the baseline requirements for the *second-generation (2G)* follow-up solution, but instead of having just a handful of countries covered, the plan now was to implement it on a global scale.

The emerging European digital cellular system was eventually named *Global System for Mobile Communications (GSM)*, although originally "GSM" referred to the French name of *Groupe Spécial Mobile*, which was the original name of the Pan-European committee that was working on the definition of the new standard. The Nordic countries were strong drivers of the GSM revolution, thanks to their comprehensive experience with NMT, and hence it was no surprise that the first ever GSM phone call occurred in Finland in 1991.

The in-built cross-country *roaming* support gave birth to another, very useful feature—a universally functioning phone numbering scheme. In pure GSM, you could call any number in the world with a full *country code–area code–actual number* sequence by starting this combined number with a "+"-sign, and it would always work correctly, independent of the country you made the call from. Therefore, if you created all of your phone book entries in this format and then went on to use your phone in another country, there was never any need to specifically figure out "how do I call home from country X".

This may sound like a small thing, but it really simplified the basic problem that was prevalent when making calls from land-line phones during traveling, as every country tends to have their own logic and prefixes for international calls.

Add to that the specific procedures needed to be able to use the phone system of the hotel you happen to reside in, and there was a lot of space for added frustration when trying to initiate a call.

Unfortunately, a handful of countries originally managed to break this beautiful, universal approach by mandating laws that required using an operator code in front of the number, in the questionable name of providing "customer choice".

The improved channel capacity was by no means the only benefit resulting from working entirely in the digital domain: continuous digital data connection between the base station and the handset provides real-time status of the quality of the received call at the base station end, which is then transmitted back to the handset and used to dynamically control its transmission power. This on-demand power management greatly increases the achievable talk times in good connectivity situations and reduces the overall interference level of the system.

By moving into the digital domain, the network also became more resistant against many common forms interference: as long as the connection between the handset and the base station was over a certain threshold, the quality of the call was excellent.

Another truly novel addition provided by digital networks was the inbuilt provision to send short textual messages between users. This *Short Message Service (SMS)* feature was added to the GSM standard as an afterthought when there was a "free" airtime slot available in the GSM *Control Channel* protocol. Most of the very first GSM phones did not immediately support SMS, but as it became wildly popular on those devices that did support it, the existence of this additional feature became an important decision factor while purchasing a new handset. Therefore, every manufacturer quickly fell in line and supported it.

Chapter 10: *Internet in Your Pocket* has further discussion of this feature.

Perhaps the most brilliant addition that was inherent to the GSM standard was the separation of the *caller identity* from the actual handset. This was accomplished through a special *Subscriber Identity Module (SIM)*, a tiny *smart card* with internal processor and memory, which the user could plug into any compatible device, thus activating her existing phone number on the device in question. The SIM card also had local memory for storing a simple address book, making it possible to have all your favorite numbers available immediately on your new device.

Significant design modifications were also done on the network side: the communications interfaces between different *network elements* that were needed for implementing a base station were standardized, which made it possible for the operators to mix and match individual components from any manufacturer of GSM-compatible equipment. This removed the chance of a *vendor lockup* due to proprietary components. Having an interchangeable standard like this was a great driver for intense competition, pushing the leading network equipment companies like *Ericsson*, *Nokia*, *Lucent* and *Alcatel* to constantly improve their products, making them ever more versatile and cheaper in the process.

Thanks to the recent advances in computing power, many of these network elements have now become *virtual*, existing only as software modules in a larger computing setup inside the base station.

With all these features, cross-compatible GSM networks spread like a wildfire across the world, and it became a truly unifying standard on a global scale, the only major exception being the United States: the history of this notable exception is explained in Chapter 9: *The American Way*.

The creation of a rapidly expanding, globally homogeneous GSM market was a boon for equipment and handset manufacturers.

As discussed in Chapter 4: *The Golden Age of Wireless*, when new technology becomes cheap enough for mass acceptance on the marketplace, the demand hits a *hockey stick curve,* and those companies who happen to ride the wave reap huge benefits.

For *Nokia*, the jackpot started with the launch of *Nokia 2110* in 1994. The planned introduction of this new device happened at the time when the technology in use for implementing handsets was about to be replaced by a new, more power-efficient generation of microprocessors. *Nokia* was originally developing a version based on the previous microprocessor generation, but thanks to visionary management, they accepted a small delay for the launch of the new product. This gave enough time for *Nokia's* Research & Development (R&D) organization to upgrade the *2110* to use the very newest and most power-efficient microprocessor technology.

The result of this change was the smallest GSM phone on the market, with battery life that was much better than the competition. *Nokia* had already come out with the first ever mass-market GSM phone, *Nokia 1011*, in 1992, so there was plenty of solid real-life experience on which to build the new version, and utilizing the newest microprocessor and *Liquid Crystal Display (LCD)* technology provided an effective base for creating a simple and intuitive *user interface (UI)* for the new device. Both *Nokia 1011* and *Nokia 2110* had full support for the SMS text messaging feature of the GSM network, but the implementation of SMS on *2110* made it very easy to use. As a result, the huge popularity of *2110* forced every other manufacturer to include SMS capabilities into their devices as well.

The effect of this combination of unparalleled features and optimal timing was enormous for *Nokia*: the original sales plan for the *2100* series was 400,000 units, but at the end of its lifespan, with all of its specific adaptations to different operating frequencies, the *2100* series had sold over 20 million units.

If a company does the profitability calculations for a new product based on $N$ number of sales and ends up selling $50 \times N$, the actual profit margin simply goes through the roof, because the unit cost of the required internal components plunges when the order volume skyrockets. Even though some of the radio interface components necessarily had to be unique to each *21XX* version, the bulk of the actual phone platform remained the same, hence using the same components over and over again. The increased order volumes for these components therefore rapidly pull down the overall *Bill of Materials (BOM)* cost of a device.

As a result, the *2100* series gave *Nokia* a huge cash boost and propelled it into the position of a leading handset manufacturer for years to come.

On a personal level, my then British employer switched my analog mobile phone to *Nokia 2110i*, and the difference in call quality, talk time and call reliability compared with my earlier analog phone was simply amazing. I remember sitting stationary in a traffic jam on the M11 motorway in London, having a conference call with my boss in New York and a co-worker in Tokyo—both of them came across crystal clear, much better actually than with my landline phone at work.

The way that conference call seemed to shrink the world felt like magic, just like the notorious Arthur C. Clarke's *Third Law* states:

Any sufficiently advanced technology is indistinguishable from magic.

Talking about Clarke, he made this ambitious prediction in 1958:

[There will be a] personal transceiver, so small and compact that every man carries one.

The time will come when we will be able to call a person anywhere on Earth merely by dialing a number.

Clarke went as far as predicting that these devices would also include a navigation function, so that:

no one need ever again be lost.

Although Clarke failed to predict the many early navigation mishaps with *Apple Maps*, it is still safe to say that his vision of the mobile future was spot on. Clarke also assumed that this development would become reality around the mid-1980s, which again exactly matched the real-life developments.

As discussed in Chapter 2: *"It's of No Use Whatsoever"*, Tesla had made his own clairvoyant prediction of wireless communications in 1926, but by also providing a verifiable and accurate timeline, Clarke takes the cake.

*Nokia* continued to be very innovative over the years. They showed the futuristic potential of mobile devices with their *Communicator* range, which offered the first ever handsets with keyboards and larger screens, only two years after the release of *2110*, and later on, they also produced Internet-oriented touch-screen devices in their *Nokia 7000* series products. Both of these were serious predecessors of current smartphones and well ahead of their time, but unfortunately, they ended up being sidelined in the later history of *Nokia*, allowing the competition to catch on.

Despite these revolutionary steps by *Nokia*, the crown for the "mother of all smartphones" belongs to *International Business Machines' (IBM) Simon*, which was on sale for only six months in 1994–1995. *Simon* already had all the hallmarks of the current crop of smartphones, including touch screen operation, calendar, email and a handful of other inbuilt applications. It even supported sending and receiving faxes—a feature that was also one of the revered features of the *Nokia Communicator* series. But *Simon*, too, was ahead of its time, and sold only 50,000 units, with a non-subsidized price that was about $1,800 in current value.

It took five more years after *Simon's* introduction before the concept of smartphone started its march towards mass adoption, originating from the youth-oriented *i-mode* phones of *NTT DoCoMo* in Japan. The fundamental driver for this next step in wireless history was the introduction of *mobile data*, and it will be discussed further in Chapter 10: **Internet in Your Pocket**.

*Nokia* remained the absolute king of the mobile hill for almost a decade, but eventually started its slow and seemingly unavoidable downward spiral, the cause of which can't be attributed to any single reason. Instead, a combination of various, often subtle issues can now be linked to this downfall:

One basic problem was the fact that *Nokia* produced both base station technology and handsets. This caused a fundamental business conflict: thanks to this duality in production, the mighty operators of the world were the main customers for both groups of products, buying handsets and network elements in huge volumes. *Nokia* had considerable self-interest to keep these operators happy. The operators naturally knew their power, and negative feedback from these high-volume customers was often very direct and had to be taken seriously—a lost order from a big operator could mean tens of millions in lost revenue.

This self-preservation instinct choked *Nokia's* numerous attempts to step into new *value-add* areas that were even remotely perceived to be part of the operators' turf.

As a comparison, any upstart competitor with no ongoing business interests could test novel ideas on the market without the fear of irritating its existing customer base. This made it possible for a company like *Blackberry* to step in and create a global mobile email business, for which *Nokia's* R&D had perfect internal solutions years before the mere existence of *Blackberry*.

What also made the operator relationship problematic was the fact that the continuously growing customer demand of *Nokia* handsets often forced *Nokia's* sales team to make hard decisions regarding product allocations across all potential clients. On the operator side, the resulting behavior of the sales force was sometimes perceived as overly arrogant, which was not necessarily a wrong conclusion, as the *Nokia* team was well aware that they could to some extent dictate their terms quite flexibly in a situation where the demand was constantly exceeding the supply. Therefore, even when *Nokia* was fully in line with a major operator customer, the huge demand for the hottest devices and the need to "balance the pain" sometimes forced *Nokia's* hand.

Those operators who felt left on the side in terms of their desired volumes or expected discounts were not happy. I personally witnessed one CEO of a major operator openly venting his anger over the decisions *Nokia* had made in respect to the product allocations and delivery schedules for his company. And not surprisingly, when the market eventually started to balance, some of these operators had elephant memories: when other manufacturers' handsets started catching up in terms of functionality and price, these operators were more than happy to "pay back" any perceived mishandling they had experienced in the past, giving *Nokia* the cold shoulder.

A case in point was the American market, which at the turn of the century was almost 100% "Nokialand", yet only ten years later it was hard to find a single *Nokia* handset on sale in American stores.

The huge volumes and the resulting massive cashflow created another mental block in *Nokia's* management: the company was no longer just an ambitious new entrant into the wireless space, instead, it had turned into a globally recognized communications behemoth, the stock price of which was discussed in great detail daily on all business channels. The unfortunate side effect of this was that *Nokia* appeared to become quite focused on protecting its stock value and optimizing its product lines accordingly. If a new, innovative gadget couldn't project sales revenues in the hundreds of millions over a relatively short term, the related R&D resources were under threat to be directed elsewhere, as there was always an urgent resourcing bottleneck somewhere else that could be seen as being more important for short-term business goals. Hence the few attempts to break new ground that did not show indications of an imminent hockey stick growth ended up being starved of resources.

One nasty side effect of this short-sightedness was the fact that *Nokia's* R&D lost several key persons who grew tired of enthusiastically working round the clock on these new leading-edge products for a year or two, just to end up closing the product program after yet another "Night of the Long Knives" had redirected the project's R&D team somewhere else.

Some persons went through three or four of these "end of program" parties, during which the existing wax models of the newly discontinued product were burned, until finally deciding to leave *Nokia* for greener pastures elsewhere.

There were some counterexamples as well: when the *third-generation (3G)* network roll-out was approaching, *Nokia* simply HAD to have a compatible handset available on the market. The team that was building it knew that even if the particular 3G handset product program was closed down, it would be reincarnated immediately in some new, slightly altered form, but with the same core team that was already familiar with the technology.

Hence, the product program was called "Kenny", based on the character from the *South Park* television series: Kenny keeps on getting killed in the most gruesome ways, but he's always back in later episodes like nothing ever happened...

The magnificent financial performance of *Nokia's* stock brought another side effect in play: hugely profitable early employer option programs which had been planned based on more modest growth expectations did not do much to help maintain a long-term business focus. When the potential income from your soon-to-be vesting option program was 10–50 times your annual salary, keeping the stock price up was a highly motivating personal driver for many. This had at least a subconscious effect on daily business decisions that had to be made.

The outcome of all this was that *Nokia* appeared to hang on to their mainstream, trusted money-making products just a little bit too long.

On the software side, *Nokia's* eventual failure with smartphones can to some extent be traced to its early, initially very successful entry to this area:

The *Symbian* operating system that was selected to be the base of the new smartphone line was originally optimized for the much more restricted hardware of *Digital Personal Assistants (DPAs)*, derived from the *EPOC* operating system of the British company *Psion Plc*. It made sense at the time, as there were no feasible alternatives around for the meek hardware that was the unavoidable norm at the time. So *EPOC* seemed like the only game in town for *Nokia's* needs.

Unfortunately, the necessary "shoehorning" approach in *EPOC* that was required in order to cope with the prevalent limitations in technology came with a lot of restrictions for the *Application Development Environment (ADE)*. *EPOC* was designed to have the smallest possible memory footprint, and this forced strict limitations for the programming model, which in turn made application development a minefield of "gotchas". The result of this was an endless source of frustration to many of the programmers entering the mobile software development scene, as most of them had personal computer or minicomputer background, both of which were much more forgiving and resource-rich environments.

Unlike software for personal computers, the software for smartphones can't be developed on the smartphones themselves. Instead, a *cross-compilation environment* is needed for the job, and the fact that the very early *Symbian* development environment was a hack based on an outdated *Microsoft Windows* C++ development environment did not help. As a C++ programmer myself, I did set up one of the earliest *Symbian* development environments, and it was very unpleasant to work with for anyone approaching the mobile environment from the direction of "traditional" programming.

But if you wanted to create applications for these budding smartphones, *Nokia's* huge market penetration made *Symbian* pretty much the only game in town. Hence, the application creators bit the bullet and programmed on—you go where the market is, and in the early days of smartphones, "mobile application market" was practically synonymous with *Symbian*.

The cross-development environment for *Symbian* did evolve relatively rapidly and became much easier to use over the years, but many of the actual systemic limitations of *Symbian* remained. As the actual hardware for wireless devices improved with leaps and bounds, these inherent limitations started to cause other problems. Issues like *real-time processing* and *platform security* caused major rewrites to the kernel of *Symbian*. These major updates forced *binary breaks* between versions, meaning that already existing applications did not run on hardware that had a newer *Symbian* version installed by default. Extra adaptation work was therefore required for every existing application that the developers wanted to make available on the newest devices. This unnecessary work hampered efforts to create a versatile *Application Ecosystem* for *Nokia's* smartphones.

Similar issues do still exist with *Apple's iOS* or *Google's Android* development environments, but on a much smaller scale than what was the norm with the numerous *Symbian* releases that were rolled out: the advances in the underlying technologies have brought the current mobile software development environments much closer to mainstream programming paradigms.

The inherent limitations of *Symbian* were well understood in the R&D department of *Nokia*, and several attempts were made to switch to *Linux*-based operating systems. Despite some promising products, like *Nokia N900*, the necessary top management support for full transition never materialized—*Nokia* kept on using *Symbian* for their mainstream smartphones due to both the perceived and the actual inertia on the market, and attempts to proceed with *Linux*-based devices were never given the kind of support that would have pushed them into the mainstream.

*Nokia* was excellent in producing cost-effective, desirable, durable and reliable *hardware*, selling hundreds of millions of copies across the world, but it never managed to make a transition into a proper, smartphone-oriented *software* company.

The final fall from grace started when Jorma Ollila, the *Nokia* CEO who had been at the helm through all the years of *Nokia's* exponential growth, had to select his successor: instead of naming a technological visionary or an experienced marketing guru as the top dog, he chose a long-term finance and legal veteran of *Nokia*, Olli-Pekka Kallasvuo.

Unfortunately, the personal image that Kallasvuo projected seemed to be in grave contrast with the leading-edge technological wizardry that *Nokia's* actual products wanted to convey to the world. If you want to be the trend-setter of a hip technology segment, the optics of your management team *do* matter, even though Kallasvuo positively tried to make fun of this apparent mismatch during several presentations. But when the reference for new competition now was *Apple* with the late Steve Jobs and his overly slick product roll-outs, even well-meaning personal denigration did not quite cut it, as *Apple's* product announcements were and still are as slick as they come.

Kallasvuo was later reported to have had his own reservations about his suitability to the top job, and whatever the actual goals for bringing Kallasvuo in might have been, the resulting impression was that *Nokia's* business focus was squarely on the next quarter instead of future mobile generations, despite many very promising R&D activities around the company.

*Nokia* had suddenly created its own *catch-22*:

Innovative new products and comprehensive platform revisions were not given the top management support they needed because their projected sales did not match short-term financial expectations. The sales projections for existing product lines were always seen as financially more viable and verifiable by the sales team, so the R&D resource allocation was done accordingly. As a result, *Nokia* lacked new, groundbreaking products, while the increasing competition and ever-cheaper off-the-shelf wireless hardware platforms rapidly eroded the margins of existing product lines.

The increased competition was primarily caused by the *commoditization* of wireless technology, which shifted the focus away from the excellence in handset technology, the area in which *Nokia* had been so good, to the excellence in the overall value-add of the devices. This fundamental shift originally started from the simple fact that the makers of microchips for wireless communications, like *Qualcomm*, started offering their own "cookbooks" on how to create mobile handsets, down to the level of *circuit board* design examples, thus making it possible to create functional wireless devices with minimal earlier experience. This nullified a lot of the benefits that *Nokia* had gained by being a pioneer in this space.

The shift towards commoditized basic wireless hardware was clearly anticipated by *Nokia's* market research, yet *Nokia* failed to divert enough focus to the value-add side of the business. Therefore, the primary source of profits remained the value that could be extracted from the actual sales of the devices, and this was eroding with lightning speed due to the new cut-throat competition.

The development in China is a good example of the major shift that happened, both in manufacturing and in R&D:

At the turn of the century, there were just a handful of budding Chinese mobile handset manufacturers, with exotic names like *Ningbo Bird*, and for actual production of handsets, *Nokia* had factories all over the world, each satisfying the needs of local markets.

Fast forward to the current situation, and pretty much everything wireless that is destined for mass-market use is manufactured in China, Vietnam or some other low-cost country by massive *contract manufacturing companies*, and some of the most innovative new smartphone products come from Chinese R&D facilities of huge companies like *Xiaomi* and *Huawei*. Even *Ningbo Bird*, a company that earlier made *pagers* and started manufacturing mobile phones only in 1999, turned out to be the largest Chinese mobile phone vendor between 2003 and 2005.

The network of factories that *Nokia* had built around the world suddenly became a costly liability in this new world of Chinese contract manufacturing that was able to undercut even the extremely streamlined production network of *Nokia*.

Hubris had also been growing alongside the seemingly limitless growth of the early years of the 21st century—the good times had been rolling just a little bit too long, and *Nokia's* position was seen as infallible by some.

As an example, when *Apple* took their first steps into the smartphone arena, *Nokia's* Executive Vice President Tero Ojanperä dismissed the budding competition as:

that fruit company from Cupertino.

After all, through their own experience, *Nokia* had already come to the conclusion that GSM data speeds offered by the first *iPhone* would not provide a good enough user experience. So why worry about a device that did not even support the newest *third-generation (3G)* standard, on which *Nokia* had just spent billions? Surely there would be enough time to get on to this new, potential cash cow when the technology matures and it can finally be projected to bring in billions in new revenues?

But the "lame" first *iPhone* was soon followed by *iPhone 3G* and a multitude of cheap *Android*-based smartphones from *Google* also emerged on the market. The worst part was that both product lines were solidly backed up with easy development ecosystems for application developers, from companies that had software development in their DNA.

And the rest is history.

*iPhone* has just had its tenth anniversary, and although everyone now admits that the first version was quite inadequate, just as *Nokia* correctly perceived at the time, it did not stop *Apple* from learning from its shortcomings and becoming the current, undisputed number one in terms of market value in smartphones.

This new threat to *Nokia's* market dominance did not go unnoticed by the board of *Nokia*, which decided that it was time for another personnel change at the top:

Stephen Elop from *Microsoft* was called in to replace Kallasvuo, and he immediately announced a groundbreaking deal with his former employer: to revive and modernize the application development ecosystem of *Nokia's* smartphones,

Nokia was going to switch to *Microsoft's* "soon to be ready" *Microsoft Phone* operating system for their future devices.

Despite the known issues with *Nokia's* smartphone production, this move stunned *Nokia's* R&D teams: a decade of *Nokia's* internal smartphone operating system R&D expertise was thrown out in one fell swoop and replaced with blind trust in the untested *Microsoft* offering instead. There wasn't a single device on sale using the *Microsoft Phone* operating system, so this was a real leap of faith.

But as there was ample room on the market for a third, competing *mobile ecosystem* to counteract *Apple* and *Google*, and *Nokia's* global marketing reach was still pretty much intact, this change alone might have been able to turn the tide for *Nokia*. *Symbian* admittedly had its teething issues, and after all, *Microsoft* was a software company, with deep understanding of the value of a consistent developer ecosystem.

All in all, this major strategy revision was a credible story that could be sold to the market, helping *Nokia* to get back to the game.

But then something completely incomprehensible happened: Elop publicly announced the end of *Nokia's* existing smartphone product line that still had several unannounced and almost ready products in the pipeline.

This was a totally illogical move at a time when the "new generation" *Microsoft Phone*-based products were not even close to being introduced to the market.

What had earlier been seen as a new, top-end of the market solution that would slowly phase out *Symbian* was suddenly announced to be the only game in town.

Business experts around the world were not surprised to see *Nokia's* *Symbian*-based handset sales take a deep dive—who would buy a product that was publicly announced to be a dead end by none other than the CEO of its manufacturer?

The results of these "crazy years" make perfect business school case study material:

In 2006, when Kallasvuo stepped in, *Nokia* had 48% of the mobile phone market.

After Elop, it was 15%.

When Elop finally sold the dwindling handset division to *Microsoft*, the market share of these "new" *Microsoft*-branded smartphones dropped into low single digits, and *Microsoft* finally killed the product line in 2016.

Still, in a way, this was a happy ending for *Nokia's* final handset misery, as *Microsoft* paid *Nokia* 7.2 billion dollars for something that it ended up scrapping just two years down the line.

The fact that *Google's Android* operating system has now become the most widely used mobile operating system bears some irony in the context of the Finnish roots of *Nokia*: *Android* is based on *Linux*, which was developed by Linus Torvards, the most famous Finn in technology. By originally choosing *Symbian* as *Nokia's* smartphone operating system, thanks to *Nokia* being the very first mass-market entrant to smartphones, *Nokia* missed the *Linux* opportunity window by just a year or two.

Still, the reluctance of *Nokia* to utilize the vast "home turf" advantage that would have come with *Linux* is one of the deep mysteries that I never understood during my time there. *Nokia* even had an early large screen tablet computer in the works, years before *iPad*, but the concept never made it past the prototype stage, mainly due to the limitations of *Symbian*. Requests to switch this potentially ground-breaking product line to use the more capable and suitable *Linux* operating system instead were repeatedly shot down by the management.

When *Nokia* finally had an active and very promising push to get cheap, *Linux*-based handsets to the market, with functional prototypes already being tested around the company, that initiative was quickly squashed after the arrival of Stephen Elop. This came as a surprise to nobody in *Nokia's* R&D, as *Linux* is the only operating system that *Microsoft* is really worried about in the personal computing space. *Linux* has already taken over as the de facto operating system for *web servers*, *supercomputers* and the ever-expanding trend of *cloud computing*, whilst *Microsoft Windows* still holds the fort for *personal computers (PCs)*.

With the demise of *Microsoft Phone*, *Linux* now rules the mobile space as well, thanks to the overwhelming market share of *Android*.

Despite this ultimate failure in smartphones, *Nokia* as a company survived, although it is now dwarfed by *Apple*—a company that, during its own period of nearly becoming a side note in computing history, was offered to *Nokia* to be bought out.

And when it comes to the business history of *Nokia*, there's another "what if" looming in the early days: just before the hockey stick curve success of *2110*, *Nokia* was offered to be taken over by *Ericsson*.

*Ericsson* declined, only to totally lose their part in the handset game, first by teaming up with *Sony*, and finally selling its remaining share of the joint venture to *Sony* in 2011.

Had either of these deals been made, the histories of both *Nokia* and *Apple* could have panned out very differently.

The story of the competition between *Nokia* and *Ericsson* has another interesting twist, showing how a single mismanagement event can radically change the later history of a company:

At the turn of the century, these two Nordic companies seemed to have equal opportunity of benefitting from the ongoing hockey stick curve. But in 2000, a factory owned by *Philips* in New Mexico, producing components for both companies, was hit by a fire that contaminated the ultra-clean manufacturing process that is necessary for microchip production. *Ericsson* accepted *Philips'* promise of restoring production within a week, whereas *Nokia's* sourcing immediately started a backup plan, searching for an alternative manufacturer of compatible components.

It was immediately obvious that there was no fully compatible replacement available, so *Nokia's* R&D changed their products to accommodate either the original *Philips* component or the near alternative that was sourced from a Japanese manufacturer.

The New Mexico factory could not get back online as promised, and as a result, *Ericsson* suffered a loss of sales worth 400 million dollars, whereas *Nokia's* sales went up by 45% on that year. This event can be traced back as the one that started

the divergence between otherwise seemingly equal competitors, eventually resulting in the downfall of *Ericsson's* handset division.

Today, both *Nokia* and *Ericsson* have lost their handset divisions, but *Nokia*
remains the world's largest company in cellular network technologies and services,
having surpassed *Ericsson* even in this area.

More interestingly, *Nokia* has now completed the full wireless technology circle
by going back to the very roots of the technology itself: after the acquisition of
*Alcatel-Lucent*, *Nokia* now owns *Bell Labs*, the original inventor of cellular
networks.

And the handset story is not totally over yet:

A company created by ex-Nokians, *HMD Global*, has licensed the use of the
name *Nokia* for their products, although it otherwise has nothing to do with *Nokia*
as a company. Like most smartphone companies, *HMD* has subcontracted the
manufacturing, and the "new *Nokias*" are being built by *Foxconn*.

These new *Nokia* smartphones run a pure version of the *Android* operating
system, something that the original *Nokia* never accepted as their choice for an
operating system. *Nokia's* long-time marketing guru, Anssi Vanjoki, even said that
using *Android* was as good as "pissing in your pants in cold weather".

In hindsight, Vanjoki could have been a better choice over Kallasvuo, and may
have been able to reverse the decline of *Nokia* even if he had been chosen as
Kallasvuo's successor instead of Elop. I watched countless marketing presentations
by Vanjoki, and I was truly impressed with his persuasion powers: he performed
very convincingly in front of the media, sometimes even managing to make them
see clear weaknesses in new products as actual strengths—to some extent, he had
similar "reality distortion field"-capabilities as Steve Jobs. But although Vanjoki
was the only one from the original "Dream Team" of *Nokia* leadership that could
have given sufficient resistance to *Apple's* slick presentations, his extravagant
management style had simply created too many enemies inside *Nokia*.

When Vanjoki resigned from *Nokia* after Elop was chosen as the CEO, the last
person from the original "dream team" that reigned over the hockey stick years of
*Nokia* was gone.

It will be interesting to see if *HMD* can pull the trick with the *Nokia* name and
the phoenix is able to rise from the ashes of the former handset division: the first
*HMD* products have just entered the market at the time of writing this, so it is yet
unknown how wet their trousers may turn out in today's cut-throat *Android*-based
smartphone market. Some early *Nokia* products were seen as indestructible and
extremely solid, whether *HMD* can match this image is yet to be seen.

As a satisfied user of a new *Nokia 5*, I wish them luck in their endeavors.

*Nokia's* handset story is another remainder that in the fast-paced cut-throat world
of global competition, a couple of bad decisions can quickly take you from being
the undisputed king of the hill of a red-hot technology to just a good, historical
name with some residual licensing potential.

# Chapter 9
# The American Way

The huge success that followed the introduction of GSM, first in Europe and then globally, caused some anxiety amongst the American technology companies. Although the United States had managed to build a comprehensive first-generation analog network, the major shift to digital suddenly seemed to happen elsewhere: the growth of GSM- based networks was phenomenal—GSM networks were available in 103 countries only five years after the first commercial implementation.

The history of the migration to digital networks turned out to be quite different in the United States, mainly due to the great success of the first-generation nationwide analog network roll-out. Although the capacity of the first-generation AMPS network was rapidly used up, it did cover virtually all of the continental United States. This was an impressive achievement, both in terms of the physical area covered and the relatively short time it had taken to reach this point.

The amount of hardware and money needed to reach a similar situation with the new digital networks would be painstakingly huge.

As the digital expansion had to be done in stages and in parallel with the existing analog network, the operators needed phones that supported both the analog AMPS and the new digital networks in the same device. Hence, specially tailored analog/ digital hybrid handsets had to be created for the US market to support this transition period.

This parallel existence of both digital and analog circuitry added complexity and cost to the handsets and initially took away the possibility of benefiting from the global economies of scale that were fueled by the digital cellular explosion internationally. As a result, it seemed like the most active pioneer of mobile technology had suddenly become trapped in its own early success.

As one of the first analog systems, AMPS suffered from all the usual problems: it was prone to interference and dropped calls, easy to eavesdrop, and was an easy target for *subscriber cloning*, in which non-encrypted subscriber information was intercepted off the air and used to initialize another phone. All subsequent calls with the cloned phone would then be charged on the account of the owner of the original phone.

© Springer International Publishing AG, part of Springer Nature 2018
P. Launiainen, *A Brief History of Everything Wireless*,
https://doi.org/10.1007/978-3-319-78910-1_9

To fight the cloning problem, some highly complicated countermeasures had to be taken. For example, the network had to be made aware of certain characteristics of the user's phone model on the actual radio interface level so that it could be distinguished from another phone from a different manufacturer that was trying to use the same, cloned identity. This was a kludge at best, sometimes causing a false alarm and cutting a totally legal customer off the network.

All this extra hassle made it look like the country that had invented the cellular phone concept seemed to be losing the game to a new European digital standard, and to make it look even worse, GSM was not a result of a single, innovative company, but instead a concoction devised by a pan-European committee, with strong contribution from government bureaucrats.

Not only was this development happening outside of the United States, but the way it came into existence appeared to be totally against the expected norm of how innovation was supposed to work in a capitalist market.

But as the AMPS networks saturated, there was no way to stick to the first-generation networks. A gradual shift to digital had to be initiated and at first, in order to maintain maximum compatibility with the existing AMPS network, a system called *Digital AMPS (D-AMPS)* was created, coexisting on the same frequency band as AMPS. This method of interleaving digital users amongst the existing analog users provided a "soft upgrade", but it was soon obvious that this could not be a lasting solution.

What happened next was extremely unusual and risky:

In order to give a boost to domestic technologies and get local companies up to speed on digital mobile communications technology, the US market eventually turned into a test bed of four different digital cellular technologies.

The first homegrown, non-AMPS compatible standard was born in the West Coast, where the local operator, *Pacific Telesis (PacTel)*, invested heavily in a budding San Diego-based company, *Qualcomm*, which had come up with a new standard proposal called *Code Division Multiple Access (CDMA)*.

CDMA used a military-style rapid *frequency-hopping* scheme, which was officially called *spread spectrum*, providing better immunity against interference, together with a theoretically more optimal utilization of the precious slice of the frequency spectrum that was available for cellular telephone use.

One particular area in which CDMA proved to be effective was dealing with long-distance *multipath propagation interference,* which refers to a situation where the transmitted signal is both received directly and reflected from various physical obstacles, like mountains, large bodies of water or high-rise buildings. This is the same effect that was the cause of the derisive *"Never the Same Color"* acronym for the American NTSC color television standard.

Another noteworthy improvement in CDMA was the more robust way to support the necessary handoff from base station to base station, effectively keeping two base stations connected to the handset until a solid handoff could be verified. This *soft handoff* feature helped in reducing the amount of dropped calls along the edges of the cells. Hence, in some crucial details of the implementation, CDMA had the potential to offer improvements over the GSM system.

The only issue was that *Qualcomm* lacked the money to provide a proof of concept, but this was soon remedied through the afore-mentioned initial investment from *PacTel*, a Californian operator struggling with the enormous growth of their customer base. Thanks to this collaboration, *Qualcomm* was able to move rapidly forward with its field testing, and the first US CDMA commercial network was opened in 1996.

CDMA was actively marketed across the world and also gained some international appeal outside of the United States—the very first commercial CDMA network was actually opened a couple of months before the *PacTel* offering by *Hutchinson Telecom* in Hong Kong.

In the years that followed, CDMA managed to gain support in a handful of locations across the world, most notably in South Korea, where the decision to select CDMA was heavily based on technopolitical reasoning in a country where the local *Chaebols*, huge industrial conglomerates, have a strong say in politics: going along with GSM would have meant that all the hardware for the Korean networks would come from existing vendors, or at least would force local manufacturers to pay heavy patent fees, so by selecting CDMA instead, the Koreans put their weight behind a new, developing standard, in which they expected to be actively participating.

In the United States, this unprecedented push of home-grown technologies initially slowed the transition to digital as compared to what was happening around the rest of the world. Having four different standards in the same geographical area resulted in a lot of overlapping, non-compatible infrastructure building, patchy networks, more costly handsets and a multitude of interoperability issues.

But as always, competition is fundamentally beneficial for technological development, and some interesting, novel features emerged from this potpourri: a good example of this application of American ingenuity was the *Integrated Digital Enhanced Network (IDEN)*, a competing network technology that was expanding the user experience in another direction by supporting so-called *push-to-talk* operation—you press a button, wait briefly for a beep indicating that an end-to-end voice channel has been opened, say what you want to say, and your voice is instantly played out at the destination handset.

No ringing, no answering. Direct connect.

This was exactly how traditional Walkie-Talkies work, except with IDEN it worked on a national scale, and proved to be a very useful feature for some particular user segments. This feature was deemed so successful that there was even an effort to update GSM to offer this kind of quick connectivity between handsets, but despite *Nokia* releasing a phone with this feature, it never caught on. This was partly due to inherent technical limitations that made the GSM version operate much more clumsily than the original IDEN version—the network logic in GSM simply was never designed to provide this kind of lightweight on-off activation for end-to-end connections, and there were too many fundamental issues that would need to be changed for optimal push-to-talk operation.

But perhaps more importantly, the operators didn't like it, as providing a "global Walkie-Talkie" feature could potentially have messed up their precious roaming revenues. The operators already had the *Short Message Service (SMS)* for text-based, instant messaging, which offered astronomical *cost per bit* data revenues, especially when the users were roaming in another country. So why add a new, even easier way of instant communications that potentially could lead to a serious loss of revenue?

I remember one particular push-to-talk system sales meeting, in which the discussion was immediately focused on the SMS revenues the operator was gaining from the text messages generated by the citizens of the country that were working abroad. Introducing a push-to-talk feature in their network could potentially decimate this lucrative cash flow, so although the company otherwise had all the newest gizmos from *Nokia* in their laboratories, push-to-talk was dead on arrival.

As with so many cases where the manufacturers and operators had tightly shared interests, it took a total outsider to finally break this global SMS goldmine, more than a decade later: *WhatsApp, Inc.*

*WhatsApp* even offers a quasi-Walkie-Talkie audio messaging facility in a form of instant voice mail feature and finally made it easy to share images between users: something that the operator-driven *Multimedia Messaging Service (MMS)* extension never managed to do.

All in all, MMS is probably the biggest dud that came out of the GSM world: whereas SMS has mostly proven to be a very reliable channel of communications even in international cross-operator cases, MMS turned out to be fragmented and extremely unreliable, and hence never really took off.

In hindsight, MMS came to existence just at the borderline between traditional, telecom-oriented, custom-tailored services and generic mobile data-oriented services, and on the wrong side of that threshold.

*WhatsApp* did not only solve the expensive international text messaging and MMS use cases: with the recent introduction of zero cost, mobile data-based audio calls, *WhatsApp* is becoming a real *one-stop-shop* threat to the very core business of existing operators.

On the audio connectivity side, this kind of pure data-based solution has already existed for quite some time through services like *Skype*, but *WhatsApp* has really taken this concept to new heights with their billion active daily users.

More detailed discussion about this purely data-driven connectivity approach follows in Chapter 10: *Internet in Your Pocket*.

At its heyday, IDEN offered great functionality for certain user segments, but eventually it could not keep up with the pace of what was happening on competing platforms, and it was finally decommissioned in 2013.

The huge economies of scale that were achieved in the GSM world were not overlooked in the United States either, and the first GSM-based network was activated in 1996. GSM was the only standard of the top four in use that could boast having inbuilt global roaming capability—a feature that was becoming more and more important for a highly developed, internationally connected nation like the United States.

There was only one snag. A big one.

The majority of GSM operators in the world built their networks on the 900 and 1800 MHz frequency bands, and the handsets were made to support them accordingly. But because these frequencies were already allocated to other uses in some countries, another pair of 850 and 1900 MHz had to be used instead.

The largest area with this kind of prior allocation was the North American market.

In the early days, multiple frequencies were costly to implement in the handsets, and therefore otherwise seemingly identical phones were sold with different active frequency bands for different markets. Hence, even though the GSM operators had made roaming agreements that made it possible for their users to keep their number while visiting another country, roaming with your own, physical phone was possible only if it supported the frequencies of the destination country.

With the next generation, dual-frequency phones that were made for their respective destination markets in order to address the huge user growth that had overwhelmed the original, single frequency band that was allocated for GSM, this cross-system functionality was not yet possible—900/1800 MHzEuropean GSM dual-frequency phone couldn't get into the US 850/1900 MHz dual-frequency networks, and vice versa.

For example, the legendary *Nokia 2110* was branded *Nokia 2190* for the version that supported the North American GSM frequencies, and it could not find a network to connect to if it was turned on pretty much anywhere else in the GSM world, despite being otherwise fully compatible with the local networks.

However, thanks to the GSM feature of separating the subscriber information from the handset, it was possible to remove the SIM card from your European/Asian/Australian phone and install it into a rented or purchased phone that used the North American frequencies. As was explained in Chapter 8: *The Hockey Stick Years*, by doing this, your number and global connectivity automatically followed you into the new handset, and if you had your phone book entries stored in the SIM memory, everything looked just like it did at home.

The "can't roam with your own phone" limitation caused by the frequency differences across the world was eventually eased, thanks to ever cheaper radio circuitry that allowed adding new supported frequencies to the handsets.

Apart from just a handful of initial 900/1900 MHz hybrid solutions, the next evolutionary step came in the form of *tri-band* devices, where the third frequency was designed to be used in intercontinental roaming cases. For example, a tri-band North American phone would support 850/1800/1900 MHz bands, allowing it to use the 1800 MHz band when roaming in the "traditional" GSM regions that were prevalent elsewhere in the world.

Further down the line, full *quad-band* frequency support became available, providing the best possible network availability on a global scale. With such a device, the only limitation for using your phone anywhere in the world was whether your home operator had a roaming agreement with the destination country, and as the often exorbitant roaming charges added nicely to the operator income, all major operators had comprehensive, worldwide agreements in place.

Sometimes you still had to specifically activate the roaming feature on your account, but many operators had it switched on by default, so your phone would work the moment you stepped out of the plane in your destination country.

Naturally, as the caller could not possibly know that the person at the other end actually happened to be in a different country, the owner of the called account had to pay the extra roaming cost for the international part. Therefore, in practice, the true hard limitation for roaming use was whether the user wanted to bear the cost of this extra convenience.

Of course, some unaware or hapless roaming users managed to ramp up astronomical phone bills during their holiday stints, providing great fodder for human interest stories in the evening news, but the bottom line is that automatic, international roaming really was a groundbreaking improvement, not only for globe-trotting businessmen but also for the occasional tourist.

Roaming support provided by GSM was an especially big deal in Europe, where you could easily visit three different countries with a total of ten or even more different operators over the course of a single day.

If your home operator has comprehensive roaming agreements, your phone is able to automatically select from several operators in the destination country, and in case one of the roaming operators has a "hole" in their coverage, your phone switches seamlessly to another operator that works better in that particular area. Hence, a person roaming in another country can potentially have superior coverage compared with any local subscribers, as they are bound to a single, local operator.

To counteract the often very high cost of roaming, the aforementioned SIM portability naturally works both ways: if you want to benefit from the lower cost that results from using a local operator instead of utilizing the roaming feature of your subscription, you can purchase a local, *pre-paid* SIM card, swap it into your existing phone, and thus give your phone a new, fully local identity.

This way, by using a local operator, your local calls are as cheap as they can be, and most importantly, your mobile data costs are just a fraction of what your roaming data would cost you, so you can now get your mandatory holiday boasting snaps into *Instagram* and *Facebook* at a much lower cost.

With the special *dual-SIM* phones, you can even keep your home number alive and accessible, while directing all data traffic through the local operator's SIM in the second slot. Usually it does require a little poking around with the configuration menus, but the potential cost savings are more than worth the hassle.

There are some positive developments that make this kind of optimization less important in certain cases: thanks to the new European Union legislation that came in force in 2017, roaming costs within the EU were slashed. This brought the European Union closer to the convenience that customers in other large geographic areas, like the United States and Brazil, had already experienced for years.

Roaming costs aside, there was also another fundamental difference regarding the billing of mobile calls: most countries separated the cellular numbers from land line numbers by new area codes, so the caller always knew that dialing a mobile number would incur higher costs.

In contrast, in some countries, most notably in the United States and Canada, mobile numbers were spread out among the existing area code space. There was no clear way of knowing whether the dialed number was a fixed line or a mobile phone, and hence all extra costs for both incoming and outgoing calls had to be borne by the owner of the mobile account.

Because of this tiny but very significant difference in the charging policy, many early cellular users in the United States kept their phones shut off, only turning them on when they themselves had to make a call, or when they got a pager message of an existing voice mail.

As a result, although the customer base grew rapidly, the air-time utilization of cellular phones in the United States initially lagged behind the rest of the world. The United States also hung on to pagers much longer than the rest of the world, where the different charging approach allowed the users to keep their mobile phones on all the time, and the existing SMS messaging of the GSM system had made paging a thing of the past.

To speed up the growth of the customer base, the American cellular operators chose to heavily promote the use of subsidized accounts. In this model, you pay either nothing or just a small sum initially for your new phone, and the operator recoups the actual cost in the form of a higher monthly fee.

To ensure that you could not take advantage of this subsidy by mixing and matching various offers from different operators, your phone was locked to work only with the SIM card that was originally assigned to you by the operator. This greatly reduced the initial cost of ownership for the customers and helped the operators grow their customer base much more rapidly than in cases where the customer would have to fork out a largish sum for an unlocked phone.

SIM lock-based subsidizing of the handset cost was also a major driver of the expansion of mobile phone use in developing countries, as paying full price for the device up front would have been too high a cost for most potential users.

The fact that the lock was in the phone and in no way limited the portability of the SIM card was hard to grasp for even some advanced users: in reality, the operator only cares about the size of the air time bill that you accumulate through the usage of your account which is assigned to your SIM card, they could not care less about the device you use to generate it with.

A good example of this lack of faith happened when my then employer *Nokia* was working on a partnership deal with another company, and I brought some of the newest *Nokia* phones as a gift to their collaboration team:

Initially, I had hard time convincing them that they really could swap in the SIM card from their existing SIM locked devices, and everything would still work just fine, without any wrath from their operator. But in the end, as usual, the lure of a new, shiny thing proved to be strong enough, and soon a bunch of happy users were yapping away with their brand-new *Nokia 8890s*, also known as *Zippos*, thanks to their smooth, cigarette lighter-like design.

*Nokia 8890* wasn't yet a tri-band GSM phone but a special, early generation hybrid "world" version, mainly aimed at high-end European and Asian users who needed to visit the United States occasionally. It had a peculiar 900/1900 MHz combination, so it was pretty well covered in the traditional GSM world, whilst the 1900 MHz frequency provided decent coverage in the urban areas of the United States.

The only major visible difference between this Nokia 8890 "world" version and the Nokia *8850* 900/1800 MHz version was the fact that the former had a retractable antenna, whereas the latter had only an internal one. This was the era when internal antenna technology was not yet on par with traditional, external antennas, and this prominent difference pretty much summed up the expected second-generation network quality in the United States during the early years of the 21st century:

The vastness of the continental United States did not help with the huge task of providing new digital coverage to replace AMPS, and the first layer of GSM networks was built with only outdoor coverage in mind.

I remember visiting Dallas, Texas in 2000 and realizing that in order to make a call with my 1900 MHz-capable phone, I had to be standing next to a window on the "correct" side of the building, which the locals called "the phone side".

Even many of the major highways did not have any coverage outside the city limits.

For anyone traveling with a roaming GSM phone, the perceived coverage difference between the United States and Europe or even between the United States and rural Asia was huge at the time. But to some extent, this was not similar to the experience that users in the United States had, as instead of having just a digital network support, most locally sold US second-generation devices still had the analog AMPS fallback mode built in. So they worked in locations where the roaming GSM phones could not find digital coverage.

Around the turn of the century, the mobile market had become truly global, and roaming had become a feature that many customers took for granted. Even in a market as large and self-sufficient as the United States, it was hard to ignore the success and interoperability that GSM enjoyed around the globe.

Despite some of its technological advantages, CDMA could not provide the kind of international interoperability that was automatically available to all GSM subscribers. The biggest misstep of CDMA was the fact that subscriber information was integrated in the phone rather than as a separate, detachable module as done in GSM. *Qualcomm* had also been very flexible in terms of adapting to local requests regarding frequencies and operator-specific tailored features, and initially this approach brought in lucrative sales contracts in some countries but it also kept the overall CDMA market very fragmented.

On the network implementation side, the fierce multi-provider competition and vast, homogeneous deployment of GSM networks ensured good adherence to system specifications, quickly weeding out any major bugs in the required base station components that provided the network coverage.

As a comparison, CDMA was somewhat lacking behind GSM in terms of overall robustness, requiring more fine tuning and maintenance to keep the networks in optimal working condition.

Although CDMA-based networks were widely installed globally, their overall market share topped at around 20%.

In an ever-expanding market, CDMA had a hard time fighting the economies of scale provided by GSM, especially in countries where CDMA had to compete with existing GSM networks over the same customer base.

A textbook example of this happened with *Vivo*, the only Brazilian CDMA-based operator: the three other major operators in Brazil, *Oi*, *TIM* and *Claro*, were using GSM and benefited from the cheaper handset prices that were possible due to the fact that virtually identical handsets could be sold across the GSM-based world. This kind of mass-market of identical devices results in lower unit prices, and the fact that *Vivo*'s handsets did not support international roaming was also a major negative point for those users who wanted or needed to travel abroad—customers in this category tend to have lots of disposable income and are the most coveted clients from the operator's point of view.

Therefore, despite already providing cellular coverage for a country of comparable size to the United States, *Vivo* decided to rip out all CDMA equipment from its network and replace them with GSM. This costly transition, which also put the company solidly on the path of an incoming *third-generation (3G)* upgrade, started in 2006 and proved to be a success: today, *Vivo* is the largest Brazilian operator.

Another quirk that resulted in from this change is the fact that due to *Vivo's* original use of the 850/1900 MHz frequency band, as compared with *Vivo's* competitors' 900/1800 MHz frequencies, Brazil is now the world's largest market that has all four GSM frequencies in parallel use. This kind of wide cross-use of available frequencies is prone to have unexpected interference issues, which the Brazilian operators had to collaboratively weed out by blacklisting some portions of the spectrum in heavily congested areas.

Although CDMA did have some features that made it perform better in certain circumstances, the sheer power of economies of scale and the non-existing roaming capability limited its international reach.

At the same time, with the globally exploding number of users, the growth of which was seemingly exponential, GSM became a victim of its own success: capacity limits were creeping in, and there was no way to squeeze more out of the existing networks.

On the other hand, the existing CDMA networks were facing exactly the same issues, especially in the United States.

Due to the exploding user base, both of these 2G systems were now facing the same problems as 1G, only on a much bigger scale, as the worldwide number of cellular subscribers was counted in billions instead of millions.

And it was also no longer just a question of voice calls: *mobile data* was the new, rapidly growing "must have" feature, and GSM was never optimized for data traffic.

GSM, being the first digital solution, was suffering from its pioneering status: years of operational practice and further research had revealed that despite its unparalleled global success, GSM was not utilizing the limited radio spectrum in the most effective way, and it was more susceptible to certain types of interference and handoff dropouts than CDMA.

Hence, when the discussions regarding the *third-generation (3G)* systems eventually started, the creator of CDMA, *Qualcomm*, was ready. They now had a handful of market-proven technical improvements that could be extremely valuable in the fight over 3G, and were keen to capitalize on their experience as well as possible.

The next revolutionary step from *mobile voice* to *mobile data* was about to start.

# Chapter 10
# Internet in Your Pocket

Another huge technological development had been happening in parallel to the ongoing digital revolution in mobile phones: the large-scale utilization of the *Internet*.

While the main focus of operators was squarely in optimizing the network performance to match the ever-growing user base for voice calls, a new feature was taking its first steps on digital networks: *mobile data connectivity*.

After all, as the networks already were based on digitized audio data, why not also provide support for generic digital data? The introduction of the *Short Message Service (SMS)* feature had already conditioned users to the possibility of instantly and reliably exchanging text-based messages while on the move, and even some clever applications had been introduced that used the SMS channel as the data connection between the handset and a backend service.

SMS was a kind of late throw-in into the GSM standard, using available bandwidth that was left "free" in the control protocol. By slapping a per-message price tag on a feature that was practically free to implement, the operators over the years have made billions of dollars from this service.

Adding a full-fledged data capability to GSM was kind of a bolted-on hindsight. The early cellular technology had focused on repeating the fixed phone line experience, only enhancing it with the new features that were possible due to the advances in electronics. This was due to the fact that for the incumbent players in the telecoms arena, voice data was seen as being very different from other data traffic.

Fundamentally though, when any kind of data has been turned into a digital form, all data transfers are just streams of bits—voice data differs from other digital data only by its expected real-time performance. Most data-oriented networks at the time did not support any kind of *Quality of Service (QoS)* separation or *traffic prioritization* between different data classes, and therefore the implementation of voice and data traffic in the 2G networks were also handled as two totally separate items.

© Springer International Publishing AG, part of Springer Nature 2018
P. Launiainen, *A Brief History of Everything Wireless*,
https://doi.org/10.1007/978-3-319-78910-1_10

During the early years of the 2G revolution, acoustic modems were the common form of connecting to the Internet. They worked by converting data traffic into audio signals of varying frequencies that were then carried over a normal phone connection and converted back to digital data at the other end of the call. From the system's point of view, you effectively had a normal point-to-point audio call as your data channel, and therefore you also paid by the connection time, not by the amount of data you consumed.

The earliest supported data implementation on GSM was *Circuit Switched Data (CSD)*, which followed the same principle in terms of cost. Similar service had already existed in some of the first-generation networks and copying it was therefore a logical step: the engineers had an existing, working model and simply copied it to the mobile world, repeating the same steps as the second-generation networks became available.

The supported speed of GSM CSD was dismal—9.6 kbps or 14.4 kbps, depending on the frequency band in use, and as mentioned, you still paid by the connection time, even if you did not transmit a single byte over the network. Therefore, CSD was a very poor deal in terms of cost per byte, given the low speeds and the high cost of mobile calls.

Despite limitations, having circuit switched data as a feature did allow general data connectivity on the go, and again, this was a major step forward for some users: being able to check your emails while on the road was a significant improvement, however slow or expensive the procedure ended up being.

By allocating more time slots per user, it was possible to extend the circuit switched speed up to 57.6 kbps through an enhanced version called *High Speed Circuit Switched Data (HSCSD)*, which was released as an upgrade. Although this allowed matching speeds with the existing acoustic modems, it was a major problem from the network's point of view, because it statically allocated up to eight time slots for a single user, severely eating up to the total available wireless capacity in a cell. Therefore, for example, in inner cities with a large number of expected simultaneous cellular users, getting full speed for your HSCSD connection was highly improbable: this kind of usage remained so rare that operators preferred to optimize their networks with the ever-increasing number of their voice customers in mind.

To really open up mobile data usage, a better way had to be developed, and *General Packet Radio Service (GPRS)*, which is often referred to as 2.5G, was the first prominent data extension to GSM. It started emerging on live networks around the year 2000.

GPRS implemented the *packet switched* data model, which was the same as what was the norm in existing fixed data networks. If you activated the GPRS mobile data feature for your mobile account, you no longer needed to allocate a dedicated data channel for the duration of your session, nor did you pay by the minute for the channel usage. Instead of a fixed, costly session where you were constantly pressed on time, you had the feel of being "always on the Internet", and you only paid for the actual amount of data being sent or received.

This was a major conceptual change, as it was now possible to have applications on the handset that were running all the time, using the data channel only when needed and without continuous user intervention.

Prior to the introduction of GPRS, if you wanted to check your email while using a circuit switched data connection, you had to specifically initiate a data call to your mail provider to synchronize your local email buffers. Due to the tedious and often costly process that in many cases just indicated that no new mail had arrived since the last synchronization session, the users only did this only a couple of times per day, potentially missing some time-sensitive mails that actually arrived between the sessions.

With GPRS and its successors, your mail application could repetitively poll the mail server on the background and download and inform you about your new mails as soon as they arrived. This fundamental paradigm shift of continuous connectivity is the core enabling feature of almost all mobile applications today, from *Facebook* to *WhatsApp*, causing the constant, asynchronous interruptions to our lives.

As they say, you gain some, you lose some.

The enhanced speed of GPRS was still based on using several time slots, but it was no longer necessary to allocate these slots for the entire duration of the connection like was the case with HSCSD. The actual, momentary data transmission speed of the connection varied dynamically, depending on the existing load in the base station: the theoretical promise of GPRS was up to 171 kbps, but this was rarely achieved in real-life situations, so compared with the speeds provided by the fixed Internet, even this best-case speed was still horrendously slow, yet it still beat the best circuit-switched HSCSD speeds up to threefold.

The next step up was the introduction of *EDGE*, which was touted as 2.75G, and released in 2003. To show how far the engineers were ready to go in order to make their alphabet soup sound snappy, EDGE stands for *Enhanced Data Rates for Global Evolution.*

The improved bandwidth that EDGE offered was implemented through updating the modulation scheme within the existing time slots. This meant that the higher speeds of EDGE did not eat up more bandwidth but were merely utilizing the existing radio resources more effectively. The maximum promised data speed was 384 kbps—still a far cry from fixed Internet speeds of the time, and as before, the speed depended on the current load on the base station.

Unfortunately, with these second-generation mobile data networks, low and often highly varying data speed was not the only issue that caused poor user experience.

First, let's look at what happens when we browse the Internet:

Common web browsing is based on *Hypertext Markup Language (HTML)* protocol, which is a simple *query-response* scheme: different parts of the web page are loaded through a series of repeated queries to the server. First, the browser loads the core portion of the page, based on the address that the user has chosen for download. It then locates any other references that reside on the page, like pictures or references to external advertisements, continuing to load each of these through additional queries.

This process of going through any remaining references on the page continues until the complete content of the originally requested page has been downloaded, and hence the overall complexity of the page defines the total amount of data that needs to be loaded. The overall time needed for this is then directly proportional to the available data transmission speed.

On top of this, each additional query-response also has to go from your device to a server somewhere in the world and back, and your actual physical distance to these servers will add delay to each roundtrip. In optimal cases, when you are near to the data source in a well-connected, usually urban setting, you may spend only 0.02 seconds extra time for each roundtrip, meaning that even a complex web page will load all its parts in a second or two. But if you are in the middle of nowhere, with poor connectivity, residing far away from the server you are connecting to, this extra time spent on each required query-response can go up to 0.5 seconds or even more.

To add to the complexity, the definition of "far" depends wholly on the *topology* of the network infrastructure you are connected on, not your actual physical distance. As an example, I'm writing this particular chapter in Manaus, which is roughly 8,000 kilometers away from London as the crow flies, but the only data connection from Manaus goes to the more industrialized part of Brazil, down south, and from Brazil, data packets going to Europe are routed first to the United States.

Therefore, according to the network trace I just performed, if I access a server in London, my data packets go first 2,800 kilometers south to Rio de Janeiro, continue 7,700 kilometers to New York, and from there another 5,500 kilometers to London. Therefore, my apparent 8,000-kilometers direct distance has now been doubled to 16,000 kilometers, and one such roundtrip consumes about 0.3–0.5 seconds of time, depending on the dynamically changing network load along the way.

This delay is solely based on the infrastructure you are connected to and is unavoidable. Part of it is due to the physical limit of the speed of light, which also limits the speed of electrons or photons making their way across the data networks, but most of it is the overhead needed to transfer these data packets across all the routers between my location and the source of the web page I am loading.

In my example, about 0.11 seconds of my roundtrip time is due to the laws of physics. That is the delay caused by the fact that my data packets need to traverse the physical network distance of 16,000 kilometers twice during each query-response roundtrip.

But more importantly, on their way between Manaus and London, my packets jump across 30 different routers, and they all need some minuscule slice of time to determine where next to send the packets.

The overall delay spent on routing and transferring the data is called *latency*, and because I was doing my example above while connected to a wired network, this was the best-case scenario. When using mobile data networks, you are facing another layer of latency, as the implementation of the air interface adds its own delay for receiving and transmitting data.

In second-generation networks, this could easily add up to one second to the processing of each single query, far exceeding the above-mentioned, unavoidable delays. Therefore, if the web page that is being loaded contains lots of subqueries, the perceived performance difference between fixed and 2G mobile networks becomes very noticeable. Hence, the further development of mobile data networks in terms of providing a better data experience was not only about data connection speed, but just as much about reducing the system-induced latency.

In terms of the available data speed, EDGE provided much more than would be needed to transfer digitized audio as pure data across the mobile network. Latency and lack of Quality of Service were still issues, but with EDGE, it was now possible to have decent quality audio connectivity purely in the data domain. The quality that was achieved with these kinds of *Voice over IP (VoIP)* services was still not on the same level as a dedicated digital voice channel, but thanks to the advances in data connectivity and achievable data speeds, the separation of audio as a special case of digital data over other types of data started to look like an unnecessary step.

Although this was not yet an issue with the 2G networks, operators had started to worry about improvements that were planned for the next generation systems: their concern was about becoming a pure *bit pipe*—if mobile data progresses to a point in which it has low latency and predictable Quality of Service, it would become possible for any greenfield company to set up VoIP services that use the operators' data networks only as a data carrier. In the worst case, operators, who had invested hundreds of millions in their network infrastructure, would wind up just transferring the data packets for these newcomers, and the customers would only look for the lowest data connectivity price, causing a race to the bottom and crashing operator revenues.

Currently, with the high-speed, high-quality fourth-generation *(4G)* data networks and eager new players like *WhatsApp*, this threat is becoming a reality, and it is interesting to see what will be the operators' response to this. Some operators have accepted this as an unavoidable next step, and are actively working on becoming the "best quality, lowest cost bit pipe" for their customers, while others, with considerable voice revenue, are looking for ways to fight this transformation.

In the current landscape of fixed price monthly packages that contain static limits for mobile call minutes, mobile data amount and the number of text messages, it is interesting to note that some operators have already included particular data traffic like *Facebook* and *WhatsApp* as unlimited add-ons. As *WhatsApp* now supports both voice and video calls, and often appears to provide much better voice quality than the up-to-the-hilt compressed mobile voice channels, giving a free pass to a competitor like this seems rather counterintuitive.

But during the early days of second-generation mobile data improvements, this potentially groundbreaking paradigm shift to all-encompassing data traffic was still a future issue: the operators had other things to worry about.

Apart from speed and latency, the screens of the handsets were a far cry from the displays that users had become used to in the fixed Internet, and to add to the insult, the available processing power and memory capacity of the handsets were tiny compared with personal computers.

But where there's demand, there will be supply.

The operators realized that enabling easy-to-use mobile data traffic would open yet another, very lucrative revenue stream, and the existing limitations of portable hardware had to be worked around to lure the customers in. Instead of supporting the existing, complex HTML protocol that was and still is the root of all Internet browsing, a much more limited set of page management commands had to be taken into use.

The first introduction of such scaled-down mobile Internet happened in Japan with *NTT DoCoMo's i-mode* service that was introduced in 1999. Despite its early mover advantage and rapid expansion to cover seventeen countries, i-mode eventually ended up failing outside of Japan. This was partially caused by *Nokia*, which did not warm to this standard that directly competed with their ongoing mobile data activities around *Wireless Application Protocol (WAP)*.

WAP was *Nokia's* first step towards the promise "to put the Internet in every pocket" made by the CEO of *Nokia*, Jorma Ollila, on the cover of the 1999 *Wired* magazine. This happened at the time when the media buzz around these kinds of grandiose promises was high, sometimes popping up in the most unexpected places: in an episode of *Friends*, Phoebe sits on a couch at *Central Perk* café, reading that particular *Wired* magazine with the cover picture of Mr. Ollila clearly in sight.

As was indicated by this kind of penetration into the Pop Culture, *Nokia* had already become the undisputed 800-pound gorilla of the mobile marketplace, and didn't hesitate to use its strong position to dictate the direction of the industry. *Nokia's* R&D was already struggling to cope with the multitude of existing differences in their mainstream markets, so adding i-mode to their product line mix would just muddy the waters even more. Only the minimum effort was therefore dedicated to handsets supporting i-mode.

As a result, any early adopters of i-mode were forced to use handsets that were coming only from Asian manufacturers, and the colorful, toy-like approach that worked in Japan and a handful of other Asian markets did not migrate too well to the rest of the world. Another blow came when the two other major operators in Japan chose WAP, and so there was no strong reason for *Nokia* to aggressively extend their activities to i-mode. *Nokia's* brief entry to i-mode compatible devices only happened six years later, with the introduction of *Nokia N70 i-mode edition* for the Singapore market. At that time, the emerging mobile data revolution was already rapidly making both i-mode and WAP obsolete.

In the meantime, WAP went through a series of incremental improvements. Small extensions, like *WAP Push*, that allowed the handset to be notified of asynchronous events like the arrival of email without actively doing a repetitive polling, truly and positively improved the perceived user experience. To pursue the possibilities for both handset sales and additional network revenues, both the device manufacturers and the operators created a massive marketing campaign around WAP.

Unfortunately, the actual services hardly ever lived up to the expectations: typically, the operator acted as a gatekeeper to all WAP content, actively trying to isolate customers from the "true" Internet. Operators wanted to extract maximum value from the content that was provided and keep it fully in their control. For the developers of content, this created an extra hurdle as they had to gain access to these *walled gardens* that had been deliberately set up by the operators.

The driver of this walled garden idea was the success that *NTT DoCoMo* had had with their i-mode standard in Japan. The value-added services in i-mode were strictly controlled by *NTT DoCoMo*, and the cash flow that *NTT DoCoMo* was raking in proved to other operators that a similar approach would be the way to go with WAP, too. Operators held the keys to their WAP gardens, which meant that every deal by content providers had to be negotiated separately instead of just "putting it on the net" for the customers to access, like is the *modus operandi* for the fixed Internet. As a result, the available WAP services tended to be very fragmented—if you changed to another operator, the set of value added services that were provided by the competing operator was very different.

Naturally some operators saw this as an additional benefit, as it sometimes stopped otherwise unhappy customers from changing their network operator.

Despite all frustrations caused by slow speeds and small screens, customers took their first baby steps with the mobile Internet, and the introduction of these often very simple services did again create a totally new revenue stream for the operators. *Ericsson*, one of the early providers of both handsets and networks, very descriptively called WAP "the catalyst of the mobile Internet", and in hindsight, both i-mode and WAP were necessary and unavoidable steps. They simply aimed at squeezing the maximum value out of the very restrictive technical limitations of the time, preparing the users for the incoming mobile data revolution that could be predicted by the expected advancements in technology in the not-so-distant future.

The WAP era also did make a handful of early application development companies momentarily very rich, although this was seldom due to the actual revenues from their products. The most prominent cases were due to the enormous hype around the "WAP boom", which was part of the wider "dot-com boom" that was engulfing almost anything that had something to do with the Internet. The hype pushed the value of the stocks of WAP-related companies sky high before crashing them down again—just like what happened with the "radio boom" in the early years of the 20th century, except now with some extravagant, often very public indulgence during the short lifespan of these companies.

Good examples of companies in this category were the Finnish WAP-application developer *Wapit*, and another Finnish mobile entertainment company *Riot-E*, both of which offered a brief but very expensive introduction to risky investing around the turn of the 21st century. *Riot-E* burned through roughly 25 million dollars in its two years of existence, and not a cent of that came from the founders of the company.

Adding WAP-capabilities to handsets was a boon for manufacturers and operators, as it accelerated the handset renewal cycle and hence kept the sales going. Despite the limited set of available WAP features, just being able to seamlessly access their email on the go was good enough reason to invest in a brand new WAP phone for many mobile handset users.

Along with these early baby steps to mobile data, the basic voice usage also kept on growing steeply, and the existing second-generation networks started to saturate under the ever-growing load.

History was repeating itself.

There was no way to cram more capacity from existing frequency bands, and the improvements in both data processing capacity and display capabilities increased the need for mobile data services. As various wireless data services started growing exponentially, mobile voice connectivity was no longer the only game in town.

The only way to move past the current limitations was another generation change, so it was time to take a look at the new developments in the manufacturers' laboratories, combine them with all the learnings gained from the field, and start pushing for a *third-generation (3G)* solution.

Although the scientists and engineers involved in the standardization process were hard-core professionals with very objective viewpoints, striving to find the best possible solution to the task at hand, there were enormous, underlying corporate interests connected to this work: whoever could get their designs incorporated into the new standard would gain huge financial benefit from the future licensing of their patents.

The standardization work tried to balance this by seeking ways to incorporate the best ideas into the new standard in novel and improved ways that would bring in the benefits with minimum cost impact in terms of existing patents, but in practice those companies with the most effective and well-funded research departments would also be able to drive this work, patent the new proposals made by their research teams, and hence ensure their share of the future licensing income.

One outcome of the US "free market" experiment was that the CDMA technology had grown to have a considerable and established user base in the United States. Despite the huge success of GSM, some theoretical improvements that could be gained by using the CDMA technology had now been proven in practice, and CDMA had its own upgrade path, called *CDMA2000 Evolution-Data Optimized (EV-DO)*, which was being pushed actively as the next generation CDMA for those operators who had CDMA in their networks.

At the same time, the holders of GSM patents were painfully aware that GSM could not offer a step up to 3G without considerable modifications: GSM, having been being early in the game, inevitably had some rough corners that were not optimized to the maximum possible level.

What was needed was a joint effort to find a solution that would avoid the kind of fragmentation that happened in the USA with the 2G networks. As a result, the maker of CDMA, *Qualcomm*, and the largest patent holders of GSM, including *Nokia* and *Ericsson*, together with other active companies in the wireless domain, created teams to work on the next generation standard.

Entire floors were booked in hotels for the herds of engineers and lawyers of the competing wireless technology companies for the duration of these negotiations, and when the dust finally settled, a proposal that combined the best parts of the competing technology directions had been forged.

The new official name of this amalgamation was *Wideband CDMA (WCDMA)*, and the major patent holders turned out to be *Qualcomm* with the largest share, followed by *Nokia, Ericsson,* and *Motorola*. It preserved the flexibility of SIM cards and the requirement of inter-operable, non-proprietary network elements of the GSM system, while adding some fundamental technological improvements offered by the CDMA technology.

Hence, despite its failure to gain major traction on a global scale, CDMA succeeded in incorporating their *air interface* to the global 3G standard. This was a major win for *Qualcomm*, helping it to grow into the prominent communications and computing hardware company it is today: at the turn of the 21st century, *Qualcomm*'s revenues were about 3 billion dollars, growing fifteen years later to over 25 billion—almost 20% annual compounded growth, and the largest money maker for *Qualcomm* in recent years has been its technology licensing business.

Just before this book went to print, Singapore-based *Broadcom Ltd.* made an unsolicited offer to buy *Qualcomm*, with a price tag of 130 billion dollars, including *Qualcomm's* existing debt. Although this was the largest ever tech takeover offer, it was rejected by the board of *Qualcomm* as being "way too low". Subsequent, even higher offer was eventually blocked by the U.S. government on national security grounds.

It just goes to show how the wireless industry has created huge value out of the effective harnessing of the electromagnetic spectrum just in the past two decades.

Similarly, although *Nokia* no longer makes handsets, according to an agreement made in 2011 and renewed in 2017, it still receives a fixed sum from every *iPhone* sold, thanks to the essential patents it holds for the cellular technology. Although the actual details are not published, with hundreds of millions of devices being sold annually, this is a major revenue stream: in 2015, the division of *Nokia* that receives related patent income from *Apple* and other licensees, indicated revenues of about billion dollars, and when the agreement was renewed in 2017, a two-billion-dollar additional one-off payment was reported.

Hence it is no wonder that new patent wars pop up all the time:

Sometimes these disputes seem to be on very different levels, though: *Apple*, a latecomer to the mobile phone business but currently making the biggest profits off it, doesn't have any significant essential patents to push. Therefore, it is not covered by the *fair, reasonable and non-discriminatory (FRAND)* agreements, where the incumbent patent owners volunteer to cross-license their standard-essential patents with lowered cost amongst the group that worked on the standard.

Instead of these patents that are essential to the actual operation of their handsets, *Apple* successfully milks their design patents of "rounded corners" and other less technical things, like the "slide-to-unlock" patent—a virtual imitation of a thousands of years old physical latch approach. Copying that basic concept ended up costing 120 million dollars for *Samsung* in 2017.

Despite its apparent simplicity compared with essential patents that many other wireless pioneers hold, this approach seems to work well: at the time of writing this, *Apple* announced that they have 250 billion dollars in the bank, and have just spent 5 billion in their new "flying saucer" headquarters in Cupertino—the most expensive office building ever.

As with every generational change in network functionality, the network hardware manufacturers try to find new, exciting use cases for the new functionality that is expected to be available in the incoming upgrade. One such promise that was much touted in relation to 3G was the support of *video calls*. This feature was prominent in all early marketing material describing what nice things were to be expected, but actual market studies soon indicated that it really was not a feature that the customers were waiting for.

Fast mobile data? Yes, please. Video calls? Not really.

Another generation switch would again be astronomically expensive, so the network manufacturers were somewhat worried about their future prospects, yet in the end, 3G turned out to be an easy sell, because for the operators the biggest issue with 2G was the severely saturated networks in densely populated areas.

So out went the futuristic Dick Tracy-video calls—the major selling point in the end for 3G was the good old voice service. It was all about matching network capacity with the growing customer demand, and despite all other advantages of 3G, enhanced mobile voice capacity turned out to be the key feature that kept the renewal cycle going. 3G also improved the encryption scheme used for the connection: the original 2G encryption method had been made practically obsolete by advances in computing power, and it was also possible to configure the 2G base stations to transparently disable the encryption.

Although there was now a clear path to interoperable 3G globally, China still threw a spanner in the works. As a single market of almost 1.5 billion customers, China is large enough to dictate their own policies as they please, and in order to support their own technology research and reduce the amount of money needed to be paid to the WCDMA patent owners, China mandated that the state-owned national operator, *China Mobile*, would be running a homegrown 3G standard. This was created with the not-so-catchy name of *Time Division-Synchronous Code Division Multiple Access (TD-SCDMA)*, and during the roll-out, it still used the standard GSM as the secondary layer for backwards compatibility.

Today, *China Mobile* is the world's largest operator by far with 850 million subscribers at the time of writing this. Hence although the TD-SCDMA standard never made it outside of China, it has a large enough user base to support very active handset and base station development.

Faster data was one of the fulfilled promises of 3G, and it has gone through a series of improvements in terms of data transmission speeds after its introduction, yet as the implementation and tuning of various features has a lot of flexibility, there is plenty of variation regarding what kind of speeds can be achieved in different operator environments.

Another important goal was to reduce the extra latency of the wireless air interface, and 3G got it down to the 0.1–0.5 second range.

And development did not stop at 3G.

The switch to *fourth-generation (4G)* systems has been progressing rapidly in the last couple of years, through a process called *Long-Term Evolution (LTE)*. This has the goal of increasing data speeds again to about five times faster in real-life situations as compared with basic 3G, and now the new generation roll-out also addressed the *upload* speed, which means that sending data like digital pictures from your smartphone is now considerably faster than with 3G.

As the LTE switch is a more gradual upgrade than what happened during the change from 2G to 3G, many operators call it *4G LTE*, which more realistically describes what is actually on offer.

*China Mobile* has also rolled out their extension to the TD-SCDMA standard, called *TD-LTE*.

For a while, another standard, called *Worldwide Interoperability for Microwave Access (WiMAX)* was pushed as a potential contender for LTE. It has its roots in the work done for the South Korean *Wireless Broadband (WiBro)* network that was launched in 2006.

To push WiMAX, *Samsung* joined forces with *Intel Corporation*, which had otherwise utterly missed the mobile revolution in its core processor business:

The new king of the "mobile hardware hill" was a British company called *ARM Holdings* in Cambridge, and as a major deviation of the norm of computer chip manufacturing, *ARM* does not manufacture its low-power/high-performance processor designs. Instead it licenses them to other manufacturers, like *Apple* and *Qualcomm*, and even to *Intel*, which finally licensed the ARM architecture in 2016. Not having to invest in expensive semiconductor fabrication plants turned out to be very profitable for *ARM*: 75% of *Arm Holdings* was bought out by the Japanese *Softbank* in 2016, with a price tag of over 30 billion dollars, meaning that the average value of this 26-year-old company grew by more than 1.5 billion dollars per year.

Despite having major players like *Samsung* and *Intel* behind WiMAX, the economies of scale were not in favor of another high-speed wireless standard that had no obvious prior user base. *Sprint Nextel* in the United States was the most prominent operator that was originally planning to have dual-mode EV-DO and WiMAX as their network infrastructure, but they eventually scrapped their plans and selected LTE instead.

WiMAX as a technology is still alive and well, however: it exists in special installations around the world, aimed at replacing fixed Internet connections.

With the increasing speeds and capacities, LTE has now progressed to a point at which it is seriously contesting fixed Internet connections. Rolling out this kind of *wireline replacement* offers a strong business case for operators: maintaining cables for fixed Internet is very costly, especially in rural setting with long cable runs and lots of thunderstorm activity. Hence, replacing a myriad of copper cables with wireless 4G LTE modems reduces projected long-term maintenance costs considerably.

Naturally, for places like emerging markets that have zero existing fixed Internet infrastructure, LTE offers a way to quickly provide fast data access to millions of potential users, meaning that a country's telecoms infrastructure can go from zero to state-of-the-art in less than a year. The potential for improved economic growth after such a leap forward is huge.

Unless your usage profile for some reason requires speeds of 100 Mbps or more, 4G LTE should be more than adequate for even those urban dwellers addicted to 24/7 video streaming, and the shift from wired to wireless has been accelerated globally after the widespread roll-out of LTE networks.

As an example, I upgraded my parents' rural fixed Internet connection to LTE-based connection, going from 500 kbps uplink + 2 Mbps downlink to 5 Mbps uplink + 30 Mbps downlink, for half the cost of the earlier fixed connection. This was one of the rare deals in which everyone was happy: the operator got rid of an error-prone, several kilometers long copper connection that had already been fried twice by a thunderstorm, and the customer saw considerable improvement in service quality, whilst still saving on the cost of the service.

One significant improvement that has vastly reduced the latency in LTE networks is the *System Architecture Evolution (SAE)* that streamlined the backend side of the networks. As a result, SAE flattened the structure of various network elements that were present in the 3G standard, reducing the overhead needed to process the transmitted and received data packets. Thanks to a low latency of under 0.1 seconds, 4G LTE is good enough even for online gaming.

And we haven't seen the end of this road yet.

As the electronic circuitry improves and processors get faster, the drive is to move to higher and higher frequency bands, allowing more bandwidth for the data.

The *fifth-generation (5G)* systems, that are just in the first deployment phase, aim to further reduce the latency and provide a ten-fold increase over the current 4G LTE network speeds. This moves the available data transmission speeds to the *gigabits per second (Gbps)* range, matching or surpassing the common fixed connection speeds at home.

As discussed in TechTalk **There's No Free Lunch**, the higher the frequency, the more room there is to modulate, and one reason for the new, higher data speeds with 5G is exactly due to the use of much higher frequency bands, which initially will cover 24–28 GHz frequencies. And for the first time, 5G is expected to allow the magic trick of supporting *full-duplex* traffic on a single frequency, therefore potentially doubling the available network capacity.

Another emerging technology that is already present in some LTE implementations is the intelligent, *multiple-input, multiple-output (MIMO)* antenna solution. This will enable dynamic *beamforming*, which means electronically changing the radiation pattern of the antenna to match the observed direction of the counterparty of a wireless connection, hence focusing the transmitted signal to the active direction only. If this trick is performed both at the base station and the handset ends, it creates a high-energy "signal bubble" at both locations, reducing interference, extending the perceived coverage area and potentially allowing a much tighter frequency reuse scheme to be used. And when the same frequencies can be reused more effectively, a single base station can handle more simultaneous connections, removing the need to install additional base stations for heavily congested areas.

This will be the standard approach for 5G: early in 2018 *Nokia* announced their new *ReefShark* 5G chipsets, which are expected to provide a throughput of up to 6 terabits (Tbps) for a single base station. To put this in perspective, this is 100 times the recorded overall data traffic at the stadium during the 2017 *Super Bowl* Football Championship game in Houston.

The MIMO technology supported by these new chipsets enables the reduction of the size of antenna structures, while at the same time they cut down the antenna setup's power consumption by two thirds as compared with conventional solutions.

Similarly, the drive in 5G is to cut down the overall power consumption of the wireless connection, therefore making it possible to connect low-power, intelligent devices directly to the existing 5G networks.

All in all, the step to 5G will be another true generational change, and it heralds the ongoing major shift from fixed Internet use to the wireless Internet: in 2017, the end users already spent 70% of their Internet-related activities on their wireless devices—an earthshaking shift that was initiated by the humble i-mode only 18 years ago.

What used to be called "mobile Internet" is becoming just "the Internet".

The world has come a long way from the early spark-gap technology, but if Marconi, Hertz or Tesla could have been transported into our time, none of them would have had any problem understanding the principles behind our latest and greatest cellular networks. Although the technology has leapfrogged, the fundamental concepts remain the same.

And all this development is still happening through conventional means: new protocols are defined, they are tested with extremely expensive, often bulky, purpose-built initial hardware, and when issues have been fixed, new purpose-built microchips are designed, paving the way to consumer-priced, portable products.

Integrating multiple functions on a single microchip allows lowering the cost and power consumption, thus pushing the technology to a price point that is needed for true mass market products.

In addition to this, when multiple wireless identities, like GPS, Wi-Fi and Bluetooth are combined into a single microchip, various interference and timing issues can be avoided by optimizing the shared logic behind these parallel operations.

Wi-Fi and Bluetooth are discussed in Chapter 11: *Home Sweet Home*.

For an emerging, vastly improved way of speeding up new developments in the utilization of the radio spectrum, see the description of *Software Defined Radio (SDR)* in TechTalk *The Holy Grail*.

Thanks to all this continuing progress in wireless technology, we consumers can seamlessly enjoy our *cat video of the day* on our smartphones, balancing the anxiety created by all of those *Instagram* shots sent by our friends and coworkers on their seemingly fabulous and entirely trouble-free holidays.

The final limiting factor is the availability of frequencies for all this use, and as discussed earlier, these frequency bands are a limited resource, strongly regulated by each country. When 3G was defined, it reused parts of the existing spectrum that was already allocated for mobile phone use, but also expanded to new, higher frequencies that were not in use yet. Part of this was due to the fact that making hardware that worked on higher frequencies used to be very costly, but had become available due to the advances in electronics.

The way these new frequencies were taken to use varied widely across the world: some countries simply allocated them for the local operators, but other countries saw the opportunity to milk this new "access to the airwaves" and opened special *spectrum auctions* for the operators to compete in.

And in many cases, these auctions turned into an incredible frenzy. For example, the government of Germany gained over 50 billion dollars off their radio spectrum auctions, whereas the United Kingdom added roughly 34 billion dollars to the government's coffers. This auction caused a serious hangover amongst the operators in Great Britain, as all this money was a sunk-in cost, separate from the funds desperately needed to purchase and deploy the required 3G infrastructure.

Unfortunately, the effects of this kind of "government leeching" can have very long-lasting negative consequences: according to a study done by Britain's *National Infrastructure Commission* in late 2016, the newest 4G LTE network in the UK was on the same level as in Albania and Peru, and at least both of them have complicated, mountainous terrain to deal with.

Hence, while the dwindling financial situations of governments can be nicely topped up by such "air sales", the citizens of these countries end up having slower deployment of the new services.

Government-mandated rules are not always bad, though: a major catalyst for intra-operator competition was the legal requirement that forced the operators to allow customers to keep their existing mobile number while changing to a competing operator. This approach has now been deployed in all major markets, and the induced potential for *churn rate*, which is the number of customers that switch to a competitor, keeps the operators on their toes, for the benefit of customers.

With available frequencies a limited resource, old services are decommissioned in order to release existing frequency bands. In the ongoing transition from analog to digital television, as was discussed in Chapter 5: *Mesmerized by the Moving Image*, the main driver is to release frequencies formerly used for television broadcasts to new uses.

As many parts of the television spectrum conveniently reside right next to existing mobile phone bands, the expectation is that these can be used to further increase the available mobile communications capacity. The auctions that happened as part of the 3G deployment have clearly shown the potential value that this might bring to the cash-strapped governments of the world.

A recent example of an auction of reallocated television frequencies happened in the United States in early 2017. Although one of the winners was the established mobile operator *T-Mobile*, the two others were *Comcast* and *Dish*, which were approaching the wireless space from a very different direction, being cable television, satellite television and Internet connectivity providers. At the time of writing this, *T-Mobile* have already announced that they will use the acquired frequencies to improve connectivity in rural areas.

This recent auction shows that the feeding frenzy seems to go on unabated—the total sum paid for these new frequency bands in the United States was almost 20 billion dollars.

Although the new frequencies will eventually expand the available wireless services, it is us customers who will finally foot the enormous auction bills, in one form or another.

Apart from the auctions, there have also been other activities aiming at benefiting from the fact that each and every television channel is not in use at every possible location and at any given time. The concept of utilizing allocated but unused television channels is called *White Spaces*, and in this context, the non-active channels can be used for unlicensed wireless Internet access, provided that the users of this technology consult a dynamically changing *geolocation database*.

This database dictates the possible frequencies that can be repurposed in a certain area at a certain time. Such databases were first defined for the United States, the United Kingdom and Canada, and those countries have also seen the first White Space-based installations go online.

The technology utilizing the White Spaces has been snappily named *Super Wi-Fi*, although it has very little to do with the actual Wi-Fi standards that are discussed in Chapter 11: ***Home Sweet Home***. Speed-wise, Super Wi-Fi is not in any ways "super" to Wi-Fi, and with its currently implemented top speed of 26 Mbps, it is also slower than 4G LTE and WiMAX, but for residents in a sparsely populated area where there are no other means to access the Internet, it provides a reasonable alternative for getting online. By using lower frequencies than the other standards, the coverage area of an individual Super Wi-Fi base station is much larger, so the deployment costs are lower than for the competing standards.

The White Spaces-concept is a great example of how a combination of modern data processing, users' geolocation information and wireless technology can be utilized to create new services that would otherwise be impossible due to existing, inflexible frequency allocations. Dynamically reusing television channels that are not in use at any given time makes it possible to maximize the utilization of our limited wireless resources, ultimately benefiting both the end users and the service providers that are happy to jump to support a niche market like this.

Our constant quest for improved bandwidth keeps on moving towards higher frequencies, while at the same time repurposing frequencies that have become obsolete due to advances in digital technology, or being "multiplexed" in terms of time and space, thanks to advances in computer technology. And as has been the case with the speed and capacity of computers, whatever improvements and advances are made, they will be utilized in full a couple of years after their initial deployment.

# Chapter 11
# Home Sweet Home

Water is something we tend to take for granted in our daily lives, even though it comes with some peculiar properties:

It is one of the few chemical substances that takes up more space in its solid form than in its liquid form, which means that ice will float, thus making the ice cubes in your drink much more pleasant to watch.

Water also breaks another common physical property by being densest at four degrees Centigrade, which is conveniently just above freezing point. This makes it possible for bodies of water to freeze from the top down and not from the bottom up—a feature that all the fish swimming happily under the ice cover would certainly appreciate, if they just stopped for a moment and thought about it.

Water molecules, which are combinations of two hydrogen atoms and one oxygen atom, as presented by their chemical formula $H_2O$, are also *electrically polarized*—they have tiny, opposite positive and negative edges, kind of like in a magnet.

A useful side effect of this polarization of water molecules was accidentally discovered by a radar engineer Percy Spencer, who had been building and installing *microwave* transmission systems for his company, *Raytheon*: in 1945, while he was experimenting with a new, high-power and small-volume microwave transmitter called *magnetron*, he noticed a tingling feeling and realized later that a peanut butter candy bar in his pocket had melted. Surprised by his observation, he experimented further by putting other foodstuff directly in front of a magnetron, including popcorn seeds, and the moment they popped, he understood the potential of this newly discovered side effect of microwaves.

A year later, *Raytheon* introduced the world's first *microwave oven*, and a novel way to cook and warm up food was born. Over the years, the microwave oven, which is in practice just a high-power microwave transmitter in a protective metallic casing, has become a standard household item, saving the day of many a bachelor worldwide by resuscitating stale slices of last night's pizza.

The inherent polarity of water molecules is behind this magic: if water molecules are put in an electromagnetic field, the existing polarity in the molecules forces

© Springer International Publishing AG, part of Springer Nature 2018
P. Launiainen, *A Brief History of Everything Wireless*,
https://doi.org/10.1007/978-3-319-78910-1_11

them to align according to the polarity of the field—positive towards negative direction of the field and vice versa. In a microwave oven, the magnetron generates a strong electromagnetic field that usually oscillates with a frequency of 2.45 GHz, meaning that the field changes its direction 2.45 billion times per second. This constantly rotating field forces the water molecules to oscillate in sync, repeatedly realigning themselves, which generates friction against adjacent molecules. And as anyone rapidly rubbing their hands against each other knows, friction causes heat.

When water molecules reside inside material like a slice of pizza, this friction generates very localized, internal heat, warming up the foodstuff around it. Therefore, any material that has some moisture in it can be heated in a microwave oven.

Glass, ceramics and plastics do not contain water molecules and only warm up through the radiation and conduction of the heat from the materials that are being heated on top of them by the microwaves. Metals, on the other hand, short-circuit the strong magnetic field and cause sparks, so if your Grandma has given you an exquisite tea cup with beautiful metal etchings on the side, don't ruin it by putting it in the microwave—use that cheap, ceramic mug you bought from *IKEA* instead for this purpose.

Living tissue is also out of bounds—give your wet cat a nice rubbing with a towel instead.

As the microwave energy penetrates deep into the material that is exposed to it, the material is not heated from the surface first, like what happens in a conventional oven that relies on conduction of heat: instead it warms up more evenly throughout, often helped by a rotating platform to counteract any fixed pattern in the electro-magnetic field that is generated by the magnetron.

This heating effect is most effective when the water molecules are free to move, meaning that the water is in liquid form, therefore warming up frozen food in a microwave oven takes much more time and should be done in burst mode, allowing minuscule water droplets to form inside the material between bursts. Most ovens have an inbuilt defrost mode for just this purpose.

Other molecules, like fats and sugars, have the same kind but weaker polar-ization, and can be heated in a microwave oven as well, but not quite as effectively as water.

One misconception and source of countless conspiracy theories comes from the notation of *microwave radiation*, which appears to have the same, ominous tone as *nuclear radiation* for some people. Using the nickname "nuking" to describe the use of a microwave oven does not help either. Luckily enough, microwaves have nothing to do with radioactivity—the only changes happening in the food during microwave heating are due to the generated heat. When the magnetron stops, the violent, forced dance of the water molecules also stops, leaving behind only the residual heat vibration that was intensified by the rapidly rotating electromagnetic field. The only changes to the food that can occur while microwaving are due to the generated extra heat: having too much of it can spoil the vitamin content of your food just as well as cooking or frying for too long would, but the fact that microwave ovens heat up food much faster usually results in an opposite outcome—more of the good stuff is present after "nuking" than after traditional cooking methods.

In any case, there is no *ionizing radiation* like when radioactive processes are present—the atomic structure of the excited molecules will stay intact.

Microwave ovens are probably the simplest application of modern radio technology: what you have is just a magnetron that transmits non-modulated signal at 2.45 GHz, focusing the beam on your soon-to-be-devoured slice of pizza. This transmitter is then encased in a metal box that is built in a way that the microwaves will not leak out of the oven.

The magnetron inside a microwave oven is essentially the same as is used in commercially available microwave radio links, as well as in the satellite transmitters that beam down your entertainment into the dish antenna of your satellite television receiver. Hence, you can thank the exact same technology for both your latest episode of *Game of Thrones* and for the popcorn you munch while watching who gets the axe this time.

The 2.45 GHz frequency was chosen for microwave ovens because it is in the middle of one of the few so-called *Industrial, Scientific and Medical (ISM)* frequency bands that are universally "free for all" to use. Contrary to many comments about microwaves, this frequency is not specifically "in tune" with water molecules: they are happy to be bounced around by almost any frequency between 1 and 100 GHz. The jointly agreed upon ISM band simply provides a convenient portion of the electromagnetic spectrum for this specific use.

Dial down the power, add suitable modulation, and the same ISM frequency band can also be used to provide a high-speed, short-range wireless connectivity:

Only 20 years ago, having a data connection to your computer used to mean that you had to plug in a fixed cable that was attached to an existing data network somewhere nearby. The problem with that naturally is that the less nerdy part of the population usually does not have a fixed data network readily installed in their home.

When the need to connect several computing devices started to arise, the tediousness and cost of providing wired connectivity to each of them individually became obvious. Especially when otherwise fully portable laptops started becoming the norm, you still had to remain within cable distance of your data modem in case you wanted connectivity to the Internet. This extra annoyance spoiled the main advantage of your laptop, which, due to its battery, was otherwise free to be used wherever you preferred to. Hence, it was no surprise that the work done in wireless data connectivity got an extra boost, and scientists in their laboratories started investigating methods of replacing cable connections with solution that would be at least almost as fast, but wire-free.

The globally agreed upon ISM frequency band that starts at 2.4 GHz was seen as an optimal choice for this purpose: it offered the potential of reasonably high bandwidth, together with a limited coverage area when low transmission power was used, both of which were seen as beneficial for localized data-oriented low-mobility use cases.

The initial steps towards this direction happened on a more fundamental level, though: the history of Wi-Fi can be traced back to the original University of Hawaii experiments with *ALOHANET*, which connected the Hawaiian Islands in 1971. *ALOHANET* supported the disassembling of a data stream into *data packets*, which allowed the collaborative sharing of the transmission channel between several participants. *ALOHANET* can therefore be seen as the forerunner of Wi-Fi, as well as a notable contributor to many other packet-based protocols.

*ALOHANET* had proven the functionality of the packet-based shared-medium approach in practice, and the same approach was applied to short-range implementations that would utilize the ISM band.

The first such solution, *WaveLAN*, was created by *NCR Corporation* in 1988, and it was originally designed to wirelessly connect a series of cash registers—not surprisingly, as the initials for *NCR* came from the earlier name of the company, the *National Cash Register Corporation*. Although *WaveLAN* was *NCR's* proprietary networking solution, it was licensed to several independent manufacturers, and thus became the first commercially available generic wireless networking solution for which it was possible to purchase equipment from multiple manufacturers. Its success encouraged electronics manufacturers to create dedicated *microchips* to support the *air interface*, which greatly helped to reduce the price of the technology.

Although most of the *WaveLAN* implementations were using the 2.4 GHz ISM band, a version with 900 MHz frequency was also available.

*NCR* later contributed their research to the *802.11 Wireless LAN Working Committee*, which has been driving the development of Wi-Fi protocols ever since.

The breakthrough in the theoretical work behind enhancing and speeding up Wi-Fi technology came from the research done in Australia in the 1990s, by a group led by John O'Sullivan: their original, unrelated research in *Radio Astronomy* indicated potential for technological advancements in wireless communications, and with funding from the *Commonwealth Scientific and Industrial Research Organisation (CSIRO)*, O'Sullivan's team created a set of patents that are the base of our current high-speed Wi-Fi technologies. Thanks to this groundbreaking work, O'Sullivan and his team were awarded the *European Inventor Award* in 2012, and he is seen as the "Father of Wi-Fi".

After a couple of iterations by the *802.11 Committee*, the very first standard, with the very user-friendly name of *802.11*, was released in 1997. It allowed data transmission speeds of up to 2 megabits per second (Mbps), which was still just on par with *WaveLAN*, but had a totally rewritten radio layer, and unlike the proprietary *WaveLAN*, *802.11* was an open standard. The first upgrade followed two years later, carrying an equally logical name of *802.11b*, this time with a considerably improved maximum speed of 11 Mbps.

For a detailed discussion on what these speeds mean in practice, see TechTalk *Size Matters*.

Early implementations suffered from interoperability issues, and in 1999, a set of companies that were active in the wireless area created the *Wireless Ethernet Compatibility Alliance (WECA)* to ensure compatibility between devices. Although the various incarnations of the standard keep on using the *802.11* naming scheme, the brand name for this overall connectivity technology was renamed Wi-Fi by *WECA,* which itself also changed its name to *Wi-Fi Alliance* in 2002.

Although the name Wi-Fi and the related logo is now prevalent in every location offering Wi-Fi access, the earlier, more technical term Wireless LAN is often still interchangeably used to refer to solutions based on 802.11 protocols.

As with any radio technology, the Wi-Fi signal is available for anyone with a matching receiver. In most cases, the users of Wi-Fi networking do not want to transmit their information openly, and hence already the preceding *WaveLAN* version had a rudimentary encryption setup built in. The *WaveLAN* encryption, however, was very easy to break, and the makers of the original *802.11* Wi-Fi standard worked on improving this important issue, first coming out in 1997 with an encryption method boldly named *Wired Equivalent Privacy (WEP).* With WEP, in order to be able to connect to the Wi-Fi network, the user had to have access to a specific *encryption key:* just knowing the network identifier was no longer enough.

Unfortunately, the security provided by WEP proved already in 2001 to be anything but "wired equivalent": it was demonstrated that with some creative trickery, it was possible to gather enough information from an active network that would eventually help in revealing the encryption key in use—worst of all, this process could be totally automated, and in case of a very active network, could be done in just a couple of minutes.

As an answer to this security breach, a totally new, enhanced system called *Wi-Fi Protected Access (WPA)* was produced in 2003, followed by the current, slightly improved standard of *WPA2.* Just prior to completing this book, the only known way to break the actual WPA2 standard was the so-called *brute force approach,* in which every possible password is tried in sequence to see if it matches the password used on the network. Therefore, with any non-trivial and long enough password, this process would take way too long to be a feasible way of breaking the security.

But unfortunately this level of safety is no longer true: after fourteen years of secure wireless networking with WPA2, the ingenuity of hackers finally caught up with it:

First, in late 2017, it was demonstrated that due to a flaw in a supporting standard, called *802.11r,* created to help users roam between cells of a large-scale WPA2-secured Wi-Fi installation in offices, it was now possible to crack even WPA2.

This, however, was still a very special use case, and WPA2 could have survived the blow, but in early 2018, a method that allows accessing the packets inside a WPA2-secured network was released: it was still impossible to figure out the network password, but any data on the network could be made visible through clever misuse of the protocol itself.

Although the manufacturers of wireless equipment swiftly moved to update their software to counteract these newly found flaws, this incident stands as a good reminder that anything wireless is potentially susceptible to eavesdropping. Hence, multiple layers of encryption should be utilized to minimize risk. More of this later.

The world now eagerly waits for the next generation WPA, WPA3, to become mainstream. This is a major upgrade that requires changes both in Wi-Fi *Access Points* and the devices that connect to it.

In the meantime, WPA2 is still your best bet, and it should always be your minimum level of encryption.

Many Access Points also offer a simplified *Wi-Fi Protected Setup (WPS)* feature, that makes it somewhat easier to connect to a WPA2-protected network by pushing a button on the Access Point and then typing in a numeric code on the client device. This feature was created by the *Wi-Fi Alliance* in 2006 to ease the use of the network for non-technical users. Unfortunately, in 2011 it was revealed that also this approach has potential security issues, so it is best to turn this WPS feature off.

It's worthwhile to note that all Wi-Fi Access Points have a configuration selection for not showing the identification of the network. This feature has nothing to do with security, as the existence of such a network is easily detected, and according to the standard, the Access Point has to provide the name when a client device requests it.

As most of us are currently using Wi-Fi as our preferred connectivity in coffee shops, hotels, airports and even in some cities with city-wide Wi-Fi coverage, it is appropriate to discuss about security in general:

When users connect to Wi-Fi networks in coffee shops and other public places, most of the time these networks are still configured without any encryption whatsoever. This approach naturally provides the ultimate in terms of easy connectivity to the network, but remember that unless you are accessing websites that are designed to be secure by using *HTTP Secure (HTTPS)* protocol, which means that they use *https://* at the beginning of the web address instead of *http://*, *anything* you type can be followed by *anyone* who is close by, simply by recording all traffic that is being transmitted across a non-encrypted wireless network.

And this is ridiculously easy to do.

By relying on an unknown service provider in a public place, it is worth understanding that *all* of your non-encrypted traffic can also be easily and transparently recorded, rerouted and mangled at will by whoever manages the network, so before you type in any of your bank connection details, ensure that your browser clearly indicates that you are actually using a secure connection. When attempting to connect to anything of value over a public Wi-Fi network, any notifications or popups that you receive should be carefully examined and taken seriously: if you observe anything out of the ordinary when accessing sensitive websites like your bank, it is worthwhile to assess whether you *really* need to do what you are doing, right there and then, or if you can wait until you are back in your at least WPA2-secured home or workplace network.

This issue is not only about accessing websites: don't forget to ensure that your email connection is using a secure protocol as well, meaning that when you log on to your email provider, the connection is encrypted. When using web-based email services, having a *https://* address and no unusual notifications during access is enough.

I tend to test this particular potential for eavesdropping with an unsecured, dummy email account connection at times, and have noticed that surprisingly many public networks "hijack" your outgoing, non-encrypted mail connection, redirecting it through their own email handler instead. Your email still works as expected, but it may be intercepted and transparently copied by an unknown third party along the way.

Fortunately, most of the existing automatic configuration setups force encryption to email traffic by default.

Another highly dangerous pitfall when trusting a third party for your connectivity comes from the feature of the Internet called *name resolution*, which is always needed to find the actual Internet address of "someplace.com" at the time you try to connect to it. These queries happen automatically "under the hood" and they can also easily be intercepted and re-routed, with a completely fake address being returned to you.

If your browser is up to date, you should receive a warning notification if an https-encrypted, fake page is offered to you as a result of a redirected name resolution, so always be extra vigilant for unusual prompts when connecting via these free Wi-Fi networks. Any warnings regarding security, or a request to retype your username and password combination in a situation in which you normally never see such a thing is highly suspicious, and if you find yourself in a situation like this, do what you need to do someplace else.

If, however, the web page you are accessing does not use https-protocol and gets redirected due to a malicious *name server*, you have no clear way of noticing it.

There are ways to protect yourself against this kind of eavesdropping and redirecting by setting up a *Virtual Private Network (VPN)* connection, in which all of your traffic is rerouted through an encrypted channel, including your name server queries, but it often comes with its own, sometimes confusing, side effects, especially if your *VPN proxy server* resides in a different country.

Using VPN also solves the newest traffic visibility hacking issue with WPA2, but further discussion over VPN is beyond the scope of this book. The cost of using a trustworthy VPN service is just a couple of dollars per month, and may save you from a lot of headache, so it is worth looking into.

Many sites that offer free connectivity try to recoup part of the cost by spying on what the users are doing, by using the tricks explained above, as well as many others. They can be as simple as producing a sellable list of locations that you have accessed while sipping your coffee, all the way to supplying invisible tags to your browser that make it easy to follow your activities across the Internet, including information on the locations where you are connecting from. Some of this may sound fairly innocuous, but remember that whatever they are doing, you can be sure that none of those activities are in place to benefit you.

It's a hostile world out there, and if something is free, it means that YOU are the product. Think of *Facebook* and *Google*.

Wi-Fi networks were originally defined to cover only small spaces, with the confined mobility within a home or a small office in mind. The maximum transmitted power is limited to 100 *milliwatts (mW)*, which is really tiny—your smartphone may use several tens of times that power when it is communicating with the cellular network, whereas the magnetron in your microwave oven blasts out a signal that is several thousand times stronger. At best, this low Wi-Fi signal strength means a coverage area with about 100-meter radius from the Wi-Fi Access Point.

With reinforced concrete walls and floors in between, the real-life coverage is usually much less than that, as the microwave radiation gets absorbed by any blocking material. To cover large spaces like airports, this problem is handled by several Access Points that are interconnected and use the same identification, allowing the users to *roam* seamlessly from one "access bubble" to another.

Another inherent limitation of Wi-Fi comes from the original frequency allocation, as there are only a couple of available channels in the 2.4 GHz band. At best, in Japan, you have fourteen individual channels to use, but most commonly only eleven channels are legal for Wi-Fi, as the edge of the ISM band is in some special, often military use in many countries.

Some Access Points come with a fixed region definition, based on the country they are sold in, others allow the user to select the region, and thus, indirectly the number of channels available.

Eleven channels may sound like a lot for a connectivity bubble with a radius of only 100 meters, but unfortunately, as the need for speed went up, there was a price to pay: more speed means more ones and zeros per time unit, which requires wider modulation, which in turn requires more bandwidth, and therefore ends up spreading any higher speed channels over multiple original Wi-Fi channels of the 2.4 GHz band.

This limitation between higher bandwidth and the resulting channel width is discussed in more detail in TechTalk *There is No Free Lunch*.

The newer, faster Wi-Fi versions follow the tradition of cryptic naming: *802.11g* provides for 54 Mbps over 2.4 GHz band, while the newest versions of *802.11n* and *802.11ac* also use the 5 GHz band, with theoretical speeds of 450 Mbps and 1,300 Mbps respectively. The added benefit of using the 5 GHz band is that it can provide at least 23 non-overlapping channels, but the higher frequency penetrates walls and other obstacles less efficiently than the 2.4 GHz signal.

The very newest version is *802.11ad*, which adds the 60 GHz band to the mix, with multi-gigabit speeds but very poor penetration through any physical obstacles.

In most real-life use cases, the achievable top speeds are usually only half or less of the promised theoretical maximum performance.

As mentioned, some of these high-speed implementations utilize the 2.4 GHz band and need more bandwidth than a single originally assigned channel can provide. Hence, using the higher speed versions on the 2.4 GHz band in practice offers only three separate channels that do not partially overlap each other—channels 1, 6 and 11.

As pretty much every domestic Internet connection you can install these days comes with its own Wi-Fi Access Point, it is not uncommon to see several tens of simultaneous Wi-Fi networks in apartment blocks: at this very moment, while typing this in a hammock on my balcony, I can see eighteen Wi-Fi networks, of which only two are my own. Eight of the others would be strong enough for me to connect to them, if I knew the corresponding WPA2 password, and every visible network has WPA2 turned on, which is good.

This amount of radio interference cacophony may sound catastrophically bad, and in special high-bandwidth cases, for example if everyone is streaming High Definition video over Wi-Fi, it truly is bad—all these networks are competing over the same scant bandwidth and cause interference to each other. But the creators of the packet-oriented data protocol, all the way back to the *ALOHANET*, understood this potential problem and designed the standard so that it tries its utmost to accommodate as many overlapping networks and users as it can.

The core of this adaptability is based on the following basic behavior:

First, if there is no data to transmit, there is minimum interference between networks, even if they are on exactly the same or partially overlapping channels. Apart from occasional announcements of the network identifier, no data is sent.

Second, when any data needs to be transmitted, it is split into smaller packets, and the Wi-Fi equipment always checks whether the channel is free before sending the next packet of data. Thanks to this *collision detection* approach, if the channel is found to be currently in use, the system waits for a tiny randomly selected delay until it tries to send the data packet again. Therefore, even in a semi-urban Wi-Fi jungle like mine, individual data packets get constantly interleaved on overlapping networks, and in practice I don't ever notice enough degradation to cause any sort of problem, even when streaming video over Wi-Fi.

One helpful procedure that is often built into most modern Access Points also tries to alleviate the problem of the existence of multiple Wi-Fi networks in the same area: it is based on the fact that the network connection is established based on the name of the network, not on the channel number in use. Therefore, the Access Point automatically scans the available frequencies at startup, selecting the least congested channel for the active operation.

For geekier users, there are plenty of free Wi-Fi scanner applications for your smartphones, tablets and laptops that you can use to see how good or bad the wireless jungle is around you. You can then use this information to optimize the channel selection, so that you get the least interference in your particular location, but as mentioned, in most cases, modern Access Points do the selection optimally for you every time they are turned on.

The bigger issue in terms of connection quality is all the concrete and iron reinforcement in the walls and ceilings, as they reduce the signal strength noticeably the further away you move from your Access Point. Lower signal strength means lower maximum speed, so it is worthwhile to install your Access Point as near to your highest-demand wireless equipment as possible.

The theoretical maximum speed can only be reached in the best possible conditions, when no other networks are sharing or overlapping the channel in use, and there are no walls or other objects between the devices that are trying to communicate. The optimal case is a rural wooden house in the middle of nowhere, with your client device in the very same room as your Access Point.

Not a very plausible scenario for most of us.

If you add another computer to the same Wi-Fi network and both computers are trying to transfer data with maximum speed simultaneously, the speed available to each of them immediately drops below 50% of the theoretical capacity. This is due to the fact that there will be numerous occasions when both computers are trying to transfer data at the same time, causing one of them to "go out for a count" and try again a bit later. Perfect interleaving of two full-speed transmission attempts is not possible as the retries rely on a random timeout.

Add to that a neighbor or three with their own, potentially overlapping Wi-Fi networks, plus one concrete wall between you and the Access Point, and your promised 54 Mbps may easily drop down to 5 Mbps or less.

The actual connection speed is constantly altered based on the existing connection quality, and this variation is transparent to the user. But again, if you want to "go geek" on this issue, there are several monitoring applications that can be used to check the ongoing connection quality and speed.

In most common use cases, you are not likely to notice this lowered speed, as even 5 Mbps is still just about good enough to transfer *High Definition* video for one device at a time, and in terms of casually browsing the Internet, 5 Mbps is a very generous speed indeed.

At the time of writing this, the most affordable, ubiquitous Wi-Fi Access Points in the 20–30-dollar range are all using the increasingly crowded 2.4 GHz band.

Only the more expensive, "high end" Access Points utilize the 5 GHz or even 60 GHz band, which are therefore much "quieter", thanks to the absence of a multitude of neighbors on overlapping channels. As an example, in my current "hammock test setup", only two of the eighteen visible networks are using the 5 GHz band.

Therefore, if you really want to be able to get the best possible throughput in your Wi-Fi network, check the small print on the box and ensure that it says "802.11n", or if you really, really have a use case for that extra speed "802.11ac" or "802.11ad".

If you are living in a detached house with concrete walls, the chances are that your Wi-Fi environment is very "clean", and buying any of these fancier Access Points is simply a waste of money. Your biggest problem is the coverage through the walls, and for that, the lower the frequency, the better the penetration of obstacles.

The very newest versions have the additional potential of supporting *beamforming*, which was briefly discussed in Chapter 10: ***Internet in Your Pocket***. Using beamforming will reduce interference and thus also improve throughput, and it will counteract the signal dampening that is happening due to obstructions between the wireless client and the Access Point. If you really think you need beamforming, it is worth reading the reviews before you buy a supporting product, as the quality of the implementation of beamforming may vary between manufacturers.

To achieve the best results, both ends of the communication link must naturally be able to use the same standard, otherwise the Access Point simply adapts to the speed of the communicating device, so if your laptop or smartphone does not support 5 GHz Wi-Fi, having an expensive Access Point does not bring any value: before forking out your hard-earned cash for the latest and greatest Access Point, check the specs of the devices you intend to connect to it.

On the low end of the price scale, there are considerable differences in the quality of the Wi-Fi implementations on various devices, mainly due to their antenna designs and antenna placement. Even though the actual Wi-Fi radios inside competing devices are often based on the same chipsets, some of the makers of which can be tracked all the way to the *WaveLAN* days, and invariably all Wi-Fi implementations use the maximum allowed power, the perceived quality tends to vary, often significantly. Some client devices work just fine in a location in which another device can't get on the network at all, and with some Access Points, the clients show a much lower signal strength than when using a competing manufacturer's product in otherwise identical conditions. Comparing your smartphone, laptop and tablet signal strengths and overall wireless performance at the very same physical location may provide very interesting results, showing that even when all devices adhere to the same wireless standard, the actual physical implementation can have a huge effect. As an example, my older, second-generation tablet often fails to connect in hotels, whilst my latest generation smartphone shows almost full-strength connectivity at the very same location.

All wireless devices are not created equal, and engineering quality still counts, just like it did in the very beginning of the wireless era.

Although the theoretical top speed is an important factor, it's more important to understand that *the slowest connection in your entire data path limits the overall speed.* If your Internet connection is only 10 Mbps and you do not have any local high-speed data sources in your home network, it is rather pointless to fork out good money for a 1,300 Mbps Wi-Fi Access Point. In a case like this, your highest speed when accessing the Internet will always be limited to a maximum of 10 Mbps due to your actual Internet connection speed, so the cheapest class of Wi-Fi Access Points will work just fine. You should get a faster Wi-Fi Access Point only if you have some local devices and data sources that would benefit from higher speeds between them across the lightning-fast local network.

If you have real issues with high-speed devices, like *High Definition* televisions accessing streamed video, and you think they are caused by interference from other active Wi-Fi networks nearby, the good old cable connection is always the safest, least troublesome solution. 100 Mbps cabling is the minimum these days, and it is very common to have 1 Gbps connections inbuilt for devices that need a lot of bandwidth.

When using a fixed cable, there are no other users than your own devices and no interference from the guy next door, so you have safe margins for even your most demanding data transfer needs. Neither is there any potential for someone eavesdropping your data.

The most powerful, potential source of interference for Wi-Fi in homes is still the microwave oven: the magnetron inside the oven has a comparable power that is 5,000–10,000 times stronger than the Wi-Fi signal, and hence the tiniest bit of leaking radio energy can cause interference, especially if your wireless device is near to the oven and the Access Point resides much further away in your home.

If you and your neighbors drop offline every time you make a bag of popcorn with your old, trusted microwave oven that you bought second-hand from *eBay* ten years ago, you should probably think of buying a new one.

If the leakage is very bad, it not only disturbs your network traffic, but the deep-penetrating heat of microwave radiation can also cause health issues, especially if you stay right next to the oven while it is being used. Thanks to the safe design of modern microwave ovens, this kind of case is rare, but if the oven door is warped or the seals are in poor condition, enough radiation may leak out to become a problem. The most sensitive body parts are eyes and testicles, the latter of which should be good enough reason for any bachelor to go and buy a replacement for that banged-out, twice-dropped-to-the-floor microwave oven.

With all the constantly growing traffic on the ISM band, a new technological development for short range connectivity is bi-directional communication based on visible light, as will be discussed in Chapter 13: *Let There Be Light*.

But when it comes to wireless connectivity on the ISM band, which is "free for all" to utilize, Wi-Fi is not the only game in town:

Another wireless communications standard that has become quite prevalent in the past couple of years goes by the name *Bluetooth*, the first truly functional version of which was standardized in 2002.

To counteract the inevitable interference issues stemming from sharing its channels on the same band as Wi-Fi, Bluetooth borrowed the same, military style *spread spectrum* frequency-hopping approach as CDMA.

In addition to just spreading the active channel over several frequencies, Bluetooth also dynamically adapts to the existing reality around it through an approach called *Adaptive Frequency-hopping Spread Spectrum (AFH)*: if some of the frequency slots in the frequency-hopping chain of a channel are detected to be constantly in use, they are ignored and only the slots that have the least traffic are actively transmitted on.

This kind of "find the holes" approach aims at offering the best possible connectivity for Bluetooth-based devices in an otherwise crowded, dynamically changing and totally unpredictable ISM frequency band.

Originally invented by the Swedish telecom company *Ericsson*, and named after an ancient Nordic king, Harald Bluetooth, this system is aimed to be a simple "wire replacement" between a master device and assorted accessory peripherals, supporting a dedicated set of roughly 40 different *profiles*, including *Hands-Free Profile (HFP)* for connecting mobile phones to your car audio system, and *Human Interface Device Profile (HID)* for connecting wireless mice and keyboards to computers.

This *master/slave* approach is a fundamental difference as compared with Wi-Fi, which was designed to be a wireless equivalent to wired, general purpose packet data networks, in which the participating devices had roughly equal status in terms of their communications abilities.

The later versions of Bluetooth extended the profiles to cover also peripheral connectivity to master devices that provide access to generic data networks, first via *LAN Access Profile (LAP)*, which has now been superseded by *Personal Networking Profile (PAN)*.

A fundamental design aspect for Bluetooth was minimizing the power consumption of the data link so that any peripherals connected via Bluetooth could use a lightweight, portable power source. To optimize this requirement, Bluetooth-compatible devices have a wide selection of transmission power levels to choose from. This naturally directly affects the maximum connection distance.

Probably the most familiar use of Bluetooth is connecting a wireless headset to your smartphone, which was also the very first use case supported by the standard. As a side effect, it has enriched our culture with a brand-new phenomenon that has become prevalent in all public places: the acceptance of a fact that people who seemingly talk to themselves aren't necessarily all crazy.

Currently the fifth major update of Bluetooth is being rolled out, and the various versions that have been released over the years have reduced the power consumption, added new profiles and augmented the maximum data rate, which now theoretically tops around 2 Mbps. This enhanced data rate is fast enough for *lossless digital audio codecs*, improving the quality of the audio on the latest generation Bluetooth headphones and wireless speakers.

Another area that has seen improvements is inherent latency: in earlier Bluetooth versions, watching a video with Bluetooth headphones sometimes had a brief but noticeable delay between picture and sound. Not anymore.

Version 5.0 also brought a welcome enhancement to the basic point-to-point wire replacement scheme of Bluetooth by enabling two simultaneous paired connections, and the maximum achievable range is now on par with Wi-Fi, at about 100 meters.

These new version roll-outs are backwards compatible.

The inbuilt flexibility regarding transmission power that is available for Bluetooth-enabled devices means that depending on the connecting device, the maximum achievable distance may be as short as just a couple of meters. This matches Bluetooth's original "wire replacement" ideology, and provides the best energy efficiency in situations where long connection distance is not necessary.

As Bluetooth was originally designed to be a protocol for simple accessories, the *pairing* of Bluetooth master and slave devices is made as simple as possible, often needing just a button press or two on both sides of the desired connection. After the pairing is done, in many cases simply turning on the accessory will restore the link to the active master.

The initial pairing state is the most vulnerable situation with Bluetooth, so it is always best to be done in a controlled environment, away from potentially hostile crowds. As new peripherals are only accepted to be paired when the master is set in a specific mode, it is very hard to add an unknown device into the master's device list without the owner first initiating or acknowledging the process. In most use cases, the data after pairing is encrypted, so that direct eavesdropping without information gathered during the pairing process is not possible.

You should never act on a Bluetooth request that you yourself did not activate. If you get any kind of Bluetooth-related query on your smartphone or other wirelessly enabled device, for example about a message that someone wants to send you, just ignore the request by selecting *cancel* from the dialog box that is showing the request to you.

In most cases like this, it is not the incoming connection itself that is the problem: it is the data that is being sent to you that contains some nasty attachment that can compromise your device. Simply do not access anything you are not fully aware of.

It is worth noting that if you have your Bluetooth connection on with the "visibility" setting switched on, the existence of your device can be found out by anyone who scans for nearby devices. This may be a problem, for example if you leave your Bluetooth-enabled laptop in your car on a quiet parking lot. Even though you have hidden it away from sight, standard Bluetooth device scan will reveal its existence, and with a descriptive Bluetooth name like "Joe's new, shiny laptop", might prompt someone to break into your car.

If you want to be 100% safe, just turn your Bluetooth feature off when you are not actively using it. Please note that many laptops keep the Bluetooth circuitry active when they are in sleep mode, so that they can be woken up via a Bluetooth-compatible keyboard, for example. If you do not want this feature, turn it completely off in settings. You can use your own smartphone to scan for available devices and verify that none of your devices are announcing their existence unless you want them to.

Bluetooth is currently being extended towards the emerging direction of the *Internet of Things (IoT)*, which is a general label for all of the wirelessly enabled gadgets in your home, from toasters to refrigerators to simple sensor devices. Most commonly they use a special subset of Bluetooth, *Bluetooth Low Energy (BLE)*. The work around this feature subset, initially called *Bluetooth Low End Extension*, was started by *Nokia* in 2001. It was first published in 2004, and became part of the Bluetooth 4.0 standard in 2010.

As configuring a home network for restricted external access is relatively hard, many of these new generation appliances have been made *cloud-enabled*, allowing you to access them easily from your smartphone or tablet. What this means is that you implicitly extend your trust to the manufacturer of these services, as the collected data from your home, like camera images, voice commands or security information, will be copied to the *web servers* that are in the complete control of the manufacturer.

There have been also cases in which some devices are "calling home" without the explicit permission of the owner, and as the actual data that gets sent in this kind of circumstances is not known, it raises issues of privacy. Many companies have valid reasons to collect various status data for quality control and further product improvement, but any cases like this should be clearly stated in the information provided for the user. In terms of privacy, it should still always be the customer's choice whether she wants to allow this kind of communication. Creating a home *firewall* setup that prohibits this kind of unwanted traffic is usually beyond the capabilities of casual users, so the second-best choice is to purchase devices only from notable companies with a good privacy reputation and clearly described functionality.

Apart from these privacy issues, this grandiose promise of "connected homes" also brings along a lot of application-level security issues: when your not-overly intelligent but wirelessly connected toaster becomes part of your home network, hacking into it may create a vulnerability that could be exploited for attacking other devices in your home.

As these devices will often need to be connected to the global Internet in order to be truly useful, they could also be used as pawns for global hacking approaches, which has already been the case with some devices, like Wi-Fi Access Points and wireless security cameras. The exploited security holes in these devices don't necessarily cause direct problem for their owners but they can be used in globally coordinated hacking or *Distributed Denial of Service (DDoS)* attacks against companies or governments. DDoSsing means flooding the target network with hundreds of thousands or even millions of simultaneous connection requests, coming from the homes and companies of unwitting users that have had their devices hacked and turned into obedient soldiers in this electronic warfare.

The first large-scale DDoS-attach happened in 1999, when the access to the University of Minnesota's network was taken down for two days due to the continuous stream of connection requests. When the attacking machines were traced and the owners contacted, they had no idea of what had been going on.

This is the general pattern for the many DDoS attacks that have happened ever since: the attacking machines are taken over by hackers, sometimes months or even years before the actual attack and without the knowledge of their owners, and then used as a combined, orchestrated "electronic army", knocking off some target network, for whatever reasons: political, business, or "just for lulz"—to show off.

In general, just following the simplest rules of thumb with any devices you connect to your home Wi-Fi network will take you a long way in terms of securing your home turf: always change at least the default password to something else, and better yet, also change the default user name used for configuring the device.

If your device offers easy access "through the cloud", you should weigh the value of this added comfort against the fact that whatever info you can access, someone at the service provider company can access as well. Naturally, to avoid potential hacking by your neighbors, or even by some *drive-by-hackers* looking for vulnerable networks, always use encryption for your wireless networks: remember that the fundamental feature of radio waves is their universal accessibility by anyone who has the right kind of equipment, so for home and office use, at minimum, always encrypt your Wi-Fi network with WPA2 using a long, complex password, and update to WPA3 whenever it comes available.

Even though unwitting users that use default passwords or no encryption are major enablers for the hackers of the world, many of the detected security issues have also been caused by the overly lax security approach of the manufactures of these devices.

Selling hundreds of thousands of devices with identical preset usernames and passwords and not actively forcing the end users to change them is actively and unnecessarily contributing to potential security holes at users' homes. Some manufacturers have even had pre-set, non-alterable secondary username/password pairs built in, creating an invisible *backdoor* to the system, thus making it extremely easy for hackers to compromise hundreds of thousands of identical devices.

There have been several serious cases of such hacking around the world, and new cases seem to happen weekly now, all due to the sloppy quality control by the device manufacturers. Internet modems with integrated Wi-Fi Access Points appear to be favored targets, as they reside on the important threshold between all wireless devices and the Internet.

To get rid of cases like these, we simply have to demand more accountability from manufacturers: a couple of high-visibility legal cases with severe financial penalties would force the industry to take security much more seriously than they appear to do. Way too often the software for these connected appliances has been contracted from the lowest bidder, with no provision for long-term maintenance.

Finally, another short-range standard worth mentioning is *Zigbee*, which is designed for low-speed, low-power use and is used in many special cases like home automation and industrial control. Zigbee is a protocol standard that can use either ISM bands or other, nationally designated frequencies. For example, in Europe, Zigbee can use 868 MHz frequency band, United States and Australia use 915 MHz, and 784 MHz is used in China.

The first Zigbee standard was published in 2004 and it is managed by *Zigbee Alliance* that was set up in 2002. Some of the notable companies behind *Zigbee Alliance* are *Comcast*, *Huawei*, *Philips* and *Texas Instruments*.

Having a huge number of low-power, intelligent devices that occasionally need to talk to each other is a good case for deploying a *mesh network*—you can read more about this in TechTalk ***Making a Mesh***.

Zigbee and recently also Bluetooth have good support for mesh networks: the latest versions of Bluetooth have special extensions that are optimized for such use through the *MESH Profile (MESH)*. Although Zigbee was originally designed with mesh networks in mind, the fact that Bluetooth is now ubiquitous gives it some edge in this ongoing fight over the *Internet of Things (IoT)*. 5G networks are also aiming to become viable connectivity providers for some IoT cases.

Both Wi-Fi and Bluetooth have brought technically very complex but easy-to-use wireless connectivity into the hands of billions of users, and are perfect examples of the ways in which the invisible wireless connectivity is fundamentally changing and enhancing our daily lives.

Despite their apparent simplicity, please spend that little extra time to ensure that you adhere to the minimum required steps needed to make your wireless network as safe as possible: do not put ease of use before security—remember that radio waves are for free for all to receive.

# Chapter 12
# "Please Identify Yourself"

On June 23rd, 1995, a beloved, four-year-old family cat named George went missing in Sonoma County, California.

For the next six months, the worried owners distributed posters and flyers, promising a 500-dollar reward for any information that would lead to George's recovery. They also repeatedly contacted all animal shelters in the neighborhood, but George seemed to have vanished without a trace.

Years went by, and they gave up on the search.

But thirteen years later the owners got a call from *Sonoma County Animal Care and Control*, stating that the shelter had picked up a stray cat that appeared to be George.

George was now seventeen years old, which is a lot for a cat, and in pretty bad shape, suffering from *toxoplasmosis* and respiratory infection. He also had a serious case of malnutrition, weighing less than three kilograms at the time he was found—about half the weight of a healthy cat.

George had clearly been busy going through his nine available cat lives in the past thirteen years, and even his last life was now hanging on a thread: in normal circumstances, if a shelter finds an animal in such a bad shape, the chances of finding a new adoptive home are very slim. For a cat in George's condition, the next step would likely have been euthanasia.

What saved George was the fact that under the loose skin around his neck was a tiny, rice grain-sized *Radio Frequency Identification (RFID)* chip, containing a serial number that could be remotely read by a simple hand scanner, and the number in George's neck brought up the contact info of its owners from the national animal database.

As is the standard at any animal shelter, the first step at Sonoma County was to check if the newly found animal had been "chipped". This is simply done by hovering a wireless reader around the animal's neck, which is the usual location for the embedded RFID chip. George's owners had had enough foresight to get their veterinarian to install the chip, and thanks to this simple procedure, done almost two decades earlier, the cat was finally reunited with its owners.

© Springer International Publishing AG, part of Springer Nature 2018
P. Launiainen, *A Brief History of Everything Wireless*,
https://doi.org/10.1007/978-3-319-78910-1_12

According to estimates, 5–7 million stray pets ended up in various animal shelters in 2016 in the United States alone, and just 20% of them were chipped, so there's still a long way to go before all runaway pets can have the same happy ending as George.

The need to be able to remotely identify objects dates all the way back to the first half of the 20th century: *Radio Detecting and Ranging (radar)* was invented in the early 1930s, and research around this novel way of remote sensing happened simultaneously in the United States, Great Britain, France, Germany, the Netherlands, the Soviet Union, Italy and Japan.

Radar was first widely used as a key tool to support the Battle of Britain: in 1940, the British coastline was sprinkled with antenna system called the *Chain Home Network*, designed to give the *Royal Air Force (RAF)* the early warning it desperately needed in trying to make this an even fight—the German *Luftwaffe* had four times as many planes as the *RAF*, so the *RAF* was looking for any possible means to gain extra advantage under the threat of imminent invasion of the British Isles.

The information gathered by the *Chain Home Network* proved crucial in balancing the situation: by being able to deploy just the right number of fighters to just the right intercept locations, the *RAF* started evening the odds: for every lost British fighter plane, two German planes were downed.

As Germany could not ensure air superiority over the English Channel during 1940, their hope for an early invasion faded. Fortuitously for the British, the Germans did not know that despite the 2:1 downing ratio, the *RAF* was still losing planes at an alarming rate and would have eventually failed.

But the situation suddenly changed due to the poor strategy choices made by Adolf Hitler: when the first British bombing raids over German territory caused a moral hit amongst the German population despite their relatively minor effectiveness, Hitler ordered *Luftwaffe* to attack primarily with bombing raids against London. This left the radar installations and airfields out of the target list, giving just the kind of breathing space the *RAF* needed in order to start a slow recovery.

Although the *Chain Home Network* was the most powerful example of the usefulness of radar, the technology itself was known to all warring parties: Germany had their own, quite successful and technologically more advanced *Freya* radar system, which showed its power for the first time in December 1939 by detecting a fleet of 22 *RAF* bombers on their way to Wilhelmshaven. Although the planes managed to finish their bombing run before the German fighters caught up with them, only half of the bombers returned back to England without any damage, whilst just three German fighters were shot down. The accuracy needed for intercepting the returning bombers was provided by the *Freya* system.

To speed up the development of the next generation *microwave* radar, based on the newly developed *magnetron*, the British decided to team up with their American counterparts, taking their most closely held secrets to a new, joint lab at the *Massachusetts Institute of Technology (MIT)* in Cambridge. This exceptional sharing of military technology happened one year before the United States even entered the Second World War.

Microwaves offered better resolution and range, and most importantly made it possible to create radar systems that were small enough to fit even onboard the airplanes. This British–American collaboration resulted in airborne radar systems that were successfully used against surfaced German submarines and other ships.

Radar is based on short radio pulses that are sent out in a highly directional beam, and then listening to any reflections of these pulses caused by objects in that particular direction. Even though the radio waves travel at the speed of light, the extremely brief time that it takes for the reflection to bounce back can be measured and used to calculate the distance to the object, whereas the current angle of the rotating antenna indicates its direction.

Radar is especially well suited for detecting objects in the air or at sea, and some special applications even work on land: several major airports utilize *Surface Movement Radar* systems that add to airport safety by providing ground controllers with real-time situational awareness regarding the movements of airplanes along the taxiways.

During the Second World War, radar had proven to be very useful in providing an early warning system of approaching airplanes, but it had one major problem: with so many planes in the air in a dynamically changing, often chaotic combat situation, how could either side tell whether the planes detected were the enemy approaching or their own planes returning from a mission?

German pilots and radar engineers noticed that if the pilots rolled their planes on their way back, the radar signature changed ever so slightly, giving the ground crew a potential indication that the approaching planes were friendly ones. The British took this approach further, installing special *transponders* to their planes that reacted to the incoming radar pulse, transmitting back a short burst on the same frequency. As a result, a stronger than normal return signal was received at the radar station, which in turn caused an easily distinguishable indication on radar screens.

The functionality of radar is the basic principle of *remote interrogation*, and identical, although more sophisticated active transponder systems are now the norm in all commercial and military airplanes.

Current aviation radar systems have two layers:

*Primary radar* matches the functional description above, giving the distance and direction of an object, but it lacks precise information of the altitude, and has no way of actually identifying the target.

*Secondary radar* sends a special interrogation pulse, and transponders on the planes reply with a short transmission that contains the plane's identification and altitude data.

Easily identifiable rotating antennas for radar can be seen at most airports, and in a very common configuration, they have a flat, rectangular secondary radar antenna stacked on top of the larger primary radar antenna.

By joining the data from both primary and secondary radar systems, the *Air Traffic Control (ATC)* can have a complete picture of exactly who is on what altitude. Computers then keep track of the relative speeds and altitudes of all detected planes, providing automated collision warnings to the controller in case two planes get too close to each other.

Most major airports have a mandatory requirement for airplanes to have their transponders active in the airspace surrounding the airport, and in practice, there should be no legal reason ever to turn off the transponder while in flight.

This is particularly important due to the fact that commercial planes are equipped with a *Traffic Collision Avoidance System (TCAS)*, which sends its own interrogation signal that is similar to the one created by the secondary radar system, and then listens to the replies coming back from the transponders of nearby planes. In case the calculated trajectory of another plane appears to be on a collision course, TCAS automatically instructs pilots on both planes to take optimal avoidance maneuvers.

This is yet another example of how radio waves can be harnessed to provide unparalleled safety in complex, constantly varying situations.

And when these systems fail, the results can be catastrophic:

In the afternoon of September 29th, 2006, flight *1907* of *GOL Transportes Aéreos* with 154 passengers and crew took off from Manaus airport for a short three-hour flight to the capital of Brazil, Brasília. For the *GOL* crew, this was yet another routine southward flight on board a three-week old *Boeing 737–800* plane, following a familiar route that at all times has only a small amount of traffic.

A couple of hours earlier, a brand-new *Embraer Legacy* business jet had taken off on its delivery flight for an American customer from São José dos Campos airport, the home base of the manufacturer, *Embraer S.A.*

*Legacy* was heading north, towards its first refueling stop at Manaus, on a flight plan that expected it to fly over the Brasília VOR station along the route.

VOR stations were discussed in Chapter 6: *Highways in the Sky*.

Less than two hours into the flight of the *Boeing*, while flying at 37,000 feet over the jungle of Mato Grosso state, the incoming *Legacy* jet's winglet hit the middle of the left wing of the *GOL* flight, slicing off half of the wing. As a result, the *Boeing* entered an uncontrollable dive, rapidly followed by an in-flight breakup and eventual high-velocity crash into a dense area of the rainforest below.

There were no survivors.

*Legacy* also had serious damage to its wingtip and horizontal stabilizer, but it managed to do an emergency landing to a nearby *Brazilian Air Force* airbase, with all five occupants unharmed.

How could two brand-new aircraft, both equipped with the very latest TCAS systems, flying in controlled airspace with very light traffic load, manage to collide?

As usual, there were several steps that finally led to the catastrophic result:

First of all, the area over the almost uninhabited Mato Grosso jungle had very poor communications coverage, and the connection between *Legacy* and ATC was lost just prior to the accident. It was clear from the recordings that both the ATC and the *Legacy* crew tried several times to talk to each other, without success. This situation with poor communications, however, was a known issue for the almost totally uninhabited area they were passing at the time, and only partially contributed to the accident.

Aviation communications is a good example of a system that is stuck with its legacy: the frequencies in use are right above the FM radio band, from 118 to

137 MHz, with the first 10 MHz dedicated for VOR beacons, as described in Chapter 6: *Highways in the Sky*. The rest is used for designated audio channels for various parts of the ATC system, from regional control centers to airport control towers, and some individual channels are used for aviation-related government agency services.

While the planes fly according to their flight plan, they have to constantly change from frequency to frequency as per the control areas they enter and exit. For modern planes flying at 800–900 km/h, this is a very frequent occurrence, and usually the workload between the two pilots is clear: one flies, while the other handles the communications.

The channel spacing used to be 25 kHz, but with the expansion of air travel facilities in densely populated areas, Europe first switched to 8.33 kHz spacing, allowing a total of 2,280 individual channels. This extended channel approach is becoming more and more widely used around the world, with all modern aviation radios supporting the 8.33 kHz channel spacing.

The modulation used for the channel is Amplitude Modulation (AM), and the voice quality is limited by the narrow available bandwidth per channel.

Channels are used in *half-duplex* mode, and one nasty side effect of the use of AM is the fact that if two transmitters are active at the same time, they usually both block each other totally, leaving only a high-volume buzzing sound to be heard on the channel. Therefore, another source of stress for pilots is managing the timing of their announcements and queries in very busy airspaces, where the control channel is effectively in constant 100% use—it is very common to have two pilots jumping into the channel at the same time, and keeping track of when the channel is not in a middle of the two-way conversation requires good situational awareness of the ATC discussions.

Similarly, as the channel is half-duplex, one stuck transmitter will block the entire channel.

All these shortcomings make aviation communications very susceptible to system failures and deliberate attacks, but due to the huge existing equipment population, switching to any kind of new system for aviation is a multi-billion-dollar issue. Plans have been made for a digital replacement, but no deadlines have been set. As for the transponder-based secondary radar model, a new *Automatic Depend Surveillance-Broadcast (ADS-B)* system is being rolled out globally. In ADS-B, the planes continuously report their GPS-based location to base stations, and there is a provision for two-way digital communications, including the view of other ADS-B-equipped airplanes in the vicinity, which will provide TCAS-style traffic awareness to even small private planes.

You can listen to many live ATC channels at:

http://bhoew.com/atc

Approach control channels for major airports, like Atlanta, Miami or New York, are good examples of high-volume traffic.

For aviation communications over the sea, separate High Frequency (HF) radios are used, or if the planes are near to shore in a busy airspace, they relay each other's messages, as was the case with *KAL 007* and *KAL 015* in Chapter 6: *Highways in the Sky*.

HF radio use is normally not expected over land. Non-commercial planes usually do not have this expensive and rarely needed feature installed, and to ensure safety during ocean crossings, they have to rent equipment that uses some satellite-based service, like *Iridium*, which is described in Chapter 7: ***Traffic Jam Over the Equator***.

Military aviation uses the same approach as the civilian one, but with frequencies which are double the civilian ones. This kind of two-to-one frequency difference, as discussed in Chapter 8: ***The Hockey Stick Years***, allows the use of an optimized antenna that covers both frequency bands, and all military aviation radios can also tune to the civilian frequencies.

In the case of *Legacy*, the distance between its location over Mato Grosso and the ATC in Brasília was just beyond the reach of communications, and such "blank spots" are not at all uncommon.

According to the original flight plan, *Legacy* was supposed to be 1,000 feet higher at the time of the accident, but this change of altitude was never commanded to it by the ATC. There would have been ample time to order *Legacy* to change its altitude while it was overflying Brasília, but no such order was given during the period of optimal radio coverage. The most recent ATC clearance transmitted to *Legacy* had defined the altitude of 37,000 feet with a destination of Manaus, and the crew is always obligated to follow the last given instruction, independent of the original flight plan.

Therefore, there was no error committed by the crew, and the ongoing discrepancy between the original flight plan and the altitude situation during the collision was a completely normal occurrence: flight plans get amended all the time to match the existing traffic and weather conditions.

The final root cause of this fatal accident was the unfortunate fact that the transponder on *Legacy* had inadvertently been switched off prior to the accident.

There is no reason whatsoever to knowingly switch a transponder off during routine flight under air traffic control, so the assumption is that the crew had made an unfortunate mistake while dealing with the control panel of their brand-new airplane. For an unknown reason, the transponder was switched off some 50 kilometers north of Brasília, and therefore the incoming *Boeing's* TCAS system, whilst being fully functional, did not get any kind of reply from *Legacy*.

For the *Boeing's* electronic TCAS eye, the *Legacy* simply did not exist.

Most unfortunately, the inactive transponder on *Legacy* also disabled its own onboard TCAS system, leaving just a small textual note on the instrumentation indicating the absence of collision detection. Therefore, *Legacy* was not sending interrogation pulses which the transponder on the *GOL* flight would have replied to.

The crew did not notice the fact that their transponder was not operational until after the accident.

The final failure in this chain of events was the fact that the air traffic controller in Brasília for some reason did not notice or indicate to the *Legacy* crew that due to the disabled transponder, their secondary radar return dropped off while the plane was still within good communications range after passing over Brasília: the air traffic controller should have been able to see a clear indication of the missing secondary radar return from *Legacy*.

The unfortunate result of all these things going wrong at the same time was that the two planes were flying at their normal cruise speeds with opposing directions on exactly the same altitude along the same *airway*, with a relative speed of almost 1700 km/h.

That is nearly half a kilometer per second.

Visually, with that kind of relative speed, even something as big as a commercial jet plane grows from a tiny dot on the horizon to a full-sized plane in just a couple of seconds.

You turn your eyes down to inspect something on the instrument panel for a moment and you miss it.

The worst thing was that as the planes were on exactly opposite trajectories, there was no apparent horizontal movement in the field of view for either crew. As discussed in Chapter 5: *Mesmerized by the Moving Image*, the human brain is extremely good at noticing sideways movement, but a dot that remains stationary and only grows gradually larger in your field of view is much harder to detect.

The *Legacy* crew reported that they only noticed a flash of a shadow, immediately followed by the thud of the accident, but that they had no idea what had hit them. The cockpit voice recorder indicated that the first comment after the collision was "what the hell was that?"

The recordings of the discussions between the *Legacy* crew and the ATC indicate that the transponder was still functional when *Legacy* was handed over to the ATC in Brasília. During the overflying of the Brazilian capital, there would have been ample time to request the *Legacy* to change its flight level, and the light traffic situation at the time would easily have accommodated such a command to be given. Instead, the flight level change request was delayed to the point at which the communications situation was so bad that it was never received or acknowledged by the *Legacy* crew.

Together with the inexplicable shutdown of *Legacy's* transponder, this unfortunate chain of events turned a perfectly harmless everyday situation into a deadly one.

The air traffic controller in duty was later deemed to be the key person who could and should have noted the situational discrepancies and imminent conflicts of currently assigned flight levels, and hence should have acted according to procedures to avoid this accident. In 2010, he was convicted and sentenced to fourteen months in jail, but the final verdict in this case appears to be still pending an appeal at the time of writing this. Both *Legacy* pilots were also given community service sentences, somewhat controversially, as they had flawlessly followed the instructions given by the Brazilian ATC, and there was neither any reason nor any indication in cockpit recordings that they intentionally turned off the transponder on their plane.

As one of the planes was under American registry, the *National Transportation Safety Board (NTSB)* performed their own investigation. Their conclusions were that the user interface of the onboard TCAS systems should be updated so that a simple text notification of a non-operational status is not the only indication that is given to the pilots.

This is yet another example of how technology is only safe when it works as expected, and even then, its existence should never be taken for granted. To have planes within meters of each other at 37,000 feet while they both are traveling at a speed of more than 800 km/h shows how incredibly precise current radio navigation and autopilot systems have become. In most cases, such precision contributes to the safety and greatly improves the capacity of the available airspace: the recent *Reduced Vertical Separation Minima (RVSM)* requirement, which is made possible due to advances in autopilot systems and is becoming commonplace in most heavily congested airspaces in the world, allows more planes to fly safer and more economic routes, with less traffic-induced delays and cancellations.

But in the case of *GOL* flight *1907*, this precision unfortunately turned out to be lethal.

Identifying airplanes is the extreme end of remote interrogation and identification. Our everyday life is full of situations where similar technology is in use, although most of us don't really pay any attention to them:

The most ubiquitous example of this is the humble *barcode*. Thanks to the *Universal Product Code (UPC)* standard used in North America, and the later, international version of *European Article Number (EAN)*, barcodes are now an integral part of practically every product that is on sale. This technology dates all the way back to 1952, when Joe Woodland was granted a patent for it, but it really became widespread after the commercial introduction of UPC in 1974.

Barcodes are read optically with a help of a *laser-based reader*, and hence they have to be visible to the reader. *Laser*, which stands for *light amplification by stimulated emission of radiation*, is one of the great inventions of the 20th century, and its properties and use in communications is further discussed in Chapter 13: *Let There Be Light*.

Thanks to the centralized management of barcodes, any company can register universally recognizable barcodes to their products: prices start from a couple of hundred dollars, covering ten individual barcodes.

By providing a cheap way to identify objects on an individual item level, the use of barcodes has thoroughly revolutionized commerce and logistics. Paying for the contents of your shopping cart no longer requires the tedious typing of the price of every single item into the cash register. Instead, the barcode reader that drives the cash register pulls not only the price of the product from the back-office database, but also other details like the name and weight, and the system promptly adds all of this information to your receipt.

Having data on the level of an individual item means that shops can have real-time inventory tracking, as well as complex research on the demand for competing product lines, or even the effectiveness of different product placement strategies inside the store. The UPC code also simplifies the basic shelf-stocking process, as in case of most products, you only need to have the price tag on the edge of the shelf, not laboriously attached to every single item. *Electronic Price Label (EPL)* systems take this even further, as the price tag contents can be changed wirelessly without manual intervention at the actual shelves.

Through barcodes, you can catalog and track electronically virtually any object large enough to have the small label attached on it. They have simplified the automated handling of bags that travel through airports, and can even provide links that you can open with your smartphone by pointing the inbuilt camera at the code in an advertisement.

For any private use, you can self-generate barcodes without having to worry about any global EAN and UPC databases: there are several standards to choose from, and most barcode readers can adapt to any of them.

The only limitation is that you still need to be able to see and read the barcodes individually, and apart from easily controllable, automated situations like airport baggage conveyor belt systems, you usually need a human in the loop to do the scanning. The optical reading also sets limits for the reading distance.

What was needed to take the barcode revolution to the next level was a way to automatically read the product tags, even without having the tags physically visible to the reader. To achieve this, it was again necessary to build the system around the application of radio waves.

Adding electronics into the equation provides another immediate obstacle: cost.

A piece of adhesive paper with some black bars scribbled on it is by far the cheapest solution available. Some companies have even dispensed with the paper: as an example, *Marks and Spencer*, one of the largest British retailers, uses lasers not only for reading the barcodes, but also to write the actual barcode directly on the skin of avocados in order to reduce paper waste.

But when it comes to cases where the encoded data is not visually readable, only the most expensive, bulkiest products could be valuable enough to warrant the use of a self-contained transceiver system. Setups like these have been in use for decades in cases like shipping containers that carry items with a combined unit worth often in millions of dollars, but for individual, smaller and cheaper items, a simpler solution had to be devised.

The answer to the cost problem came in the form of a *passive RFID tag*.

The first such solution, which eventually led to the embedded RFID chips in cats like George, was patented in 1973 by Mario Cardullo—a year before the UPC code was adopted for use in barcodes. His invention described an electronic circuitry that was able to work without an internal battery, consisting of a very simple combination of electronic memory and surrounding management circuitry.

When a radio signal that is transmitted by the *interrogator* hits the tag's inbuilt antenna, it generates a tiny current that is extracted to activate the electronics on the tag. The electronics thereafter modify the electrical aspects of the receiving antenna according to the data in the tag's memory. This change causes subtle modulation in the *backscatter* signal that gets reflected back from the tag's antenna, and the resulting, modulated backscatter signal can then be detected by the interrogating electronics, thus reading the contents of the tag.

This backscatter-alteration method is a refined version of the "roll the wings"-approach used by the German planes in the Second World War.

What makes passive tags so cheap is the fact that the "receiver" is actually just a simple antenna, connected to a resonator that is tuned on a fixed frequency, and the "transmit" process is simply an adjustment of the electrical properties of the resonator. Hence, there is no real, active receiver or transmitter in the tag, and the number of electronic components required for this functionality is minimized.

The interrogation process of RFID tags does not need to be based only on radio waves: systems that use an *inductive* or *capacitive* approach also exist, but the radio approach provides the longest and most flexible interrogation distance.

Solutions that offer alteration of all or part of the tag's internal memory under the command of the interrogator also exist, which enable the remote alteration of the tag's content without physically replacing it, but for the majority of solutions, cheap read-only memory is all that is needed.

The advances in electronics that have led to extremely cheap, low-power, and solid-state components, have made it possible to directly harness the incoming radio energy to power the RFID circuitry. This removes the need for a battery and thus gets rid of this costly and bulky component, which, unlike other electronics on an RFID tag, inevitably has a limited life span.

Thanks to the extremely simple electronics needed on the tag and the wide deployment of RFID technology, the cost of the embedded electronics has plummeted: currently, a passive RFID tag costs about 10 cents, and the price continues to go down. This is still not useful for the very cheapest items, but becomes viable for anything that has a value of at least a few tens of dollars.

The fact that there is no need for an optical line of sight connection between the interrogator and the RFID tag is the major benefit over barcodes. Depending on the standard in use, the reading distance can vary from centimeters to several tens of meters.

Early experiments with RFID technology were in the area of automated toll collection on roads, tunnels and bridges, as well as access control in buildings. Systems like these are now in wide use in our daily lives, including the relatively recent addition of contactless payment systems and remotely readable passports.

Some of the very first tests experimented with animal identification, which led to the current procedure of "chipping"—a practice that saved George and many other pets. Several countries require a match between the pet's embedded ID and supporting vaccination records before the animal is allowed to enter the country, and hence any internationally mobile cats and dogs need to be chipped by default.

Newer radio-frequency RFID systems allow multi-reading, ranging from a minimum of about ten items up to several thousand. This brings an added flexibility to logistics, as the individual items inside a cargo container can be read in one go.

The tags do not all respond at the same time, though, as this would cause severe interference and nothing could be read. Various anti-collision methods are used, in a similar fashion to what happens with Wi-Fi, so that eventually all the tags have been sequentially interrogated, with speeds that can be as fast as 300–500 tags per second.

For consumers, the future promise of RFID in shopping is eventually removing the cash register step altogether: if all products in your shopping cart have RFID tags, you can just push your cart past the multi-scanner and pay. And if you use a contactless payment solution embedded in your smartphone or credit card, even your payment will be based on remote interrogation.

This concept is still in experimental phase around the world: the new, much-hyped, fully automated *Amazon* stores are currently using a combination of camera-driven customer tracking and specialized large-format optical barcodes.

RFID is yet another good example of how new inventions can quickly permeate through our existing processes and practices, making them easier, more reliable and faster along the way.

When the first register-less, RFID-based store opens near you, remember that all of this ingenious utilization of radio waves can be traced back to the time when German Second World War pilots noticed that rocking the wings changes the radar signature of their planes.

# Chapter 13
# Let There Be Light

The microwave absorption caused by hard rain or snowfall becomes more notable towards the end of the microwave spectrum at around 300 GHz, which corresponds to a *wavelength* of about one millimeter. Above this frequency, the behavior of the electromagnetic radiation fundamentally changes.

The concept of wavelength is discussed in TechTalk ***There is No Free Lunch***.

As the frequency of electromagnetic waves keeps on rising beyond microwaves, we first enter the infrared portion of the electromagnetic spectrum, followed by visible light, and then ultraviolet radiation.

Infrared radiation is something that is very familiar in our lives, although we probably don't think about it much—we all not only transmit it continuously but can also concretely detect large amounts of it on our skin, as any warm object gives off infrared radiation.

Despite its very different characteristics, we can modulate a signal in the infrared range just like any other electromagnetic radiation, but it no longer passes through walls. Instead, infrared radiation gets partially absorbed by the obstructing material, and then slowly travels through it by means of *conduction*, which effectively dampens away any modulation that existed in the original infrared signal.

The part of infrared radiation that does not get absorbed by an obstruction gets reflected in a similar fashion as visible light. This means that infrared radiation is confined within the walls of a room, and this is a perfect limitation for the most common usage of modulated infrared radiation that is present in practically every household: the ubiquitous *remote controller*—a device that has given us the freedom to switch channels the moment the channel we're watching is no longer keeping us interested, to the dismay of all the advertisers that have spent big money trying to catch our eye during commercial breaks.

The infrared radiation power transmitted by the remote controller is very tiny, so you won't be able to feel the heat with your finger, but if you watch the infrared *light-emitting diode (LED)* at the tip of your remote controller through the camera of your smartphone, you will see that it lights up when you press the buttons. Although the frequency of the infrared radiation is beyond the limits of your eye,

© Springer International Publishing AG, part of Springer Nature 2018
P. Launiainen, *A Brief History of Everything Wireless*,
https://doi.org/10.1007/978-3-319-78910-1_13

most camera sensors are sensitive beyond the visible light spectrum and thus make it possible to "see" this stream of infrared pulses.

The fact that most of us can't remember the last time we changed the batteries in our remote controller is an indication of how low power the transmitted signal actually is, and this minuscule signal strength poses some major issues in our heat-drenched environment: direct sunlight can create a million times stronger infrared signal than our humble remote controller.

The seemingly magical feat of still being able to flip the channel of a television on our back porch on a sunny day is achieved through two special features:

First, the transmitter and receiver are tuned on a very thin portion of the infrared spectrum, usually with a wavelength of 980 nm (billionths of a meter), which matches a frequency of about 300 terahertz (THz). This passive filtering is the reason why practically every infrared receiver is covered with a dark red, opaque piece of plastic. Thus, by physically blocking out a big chunk of unwanted infrared frequencies, we provide the first step in ensuring functional communications between the infrared transmitter and receiver.

But the most important feature that makes our remote controllers very reliable is the *active modulation* of the transmitted signal: by modulating the pulses of infrared light with a known *carrier frequency*, the receiver can use an electronic circuit called a *phase-locked loop* to fixate only on infrared pulses that match the designated carrier frequency.

There are numerous different approaches to how this modulation and pulsing of the infrared signal is formulated, but basically they all convert commands into a stream of digital bits, the meaning of which is totally device-dependent. Any single manufacturer tends to use the same encoding approach across all the multiple different devices it manufactures, but by including a specific device ID in the bit stream, the receiving devices can distinguish the commands that are directed to them only.

Sending a single command takes only a very short time, meaning that the same code is normally sent repeatedly several tens of times per second. Due to the digital format and the usual sequential transmission of the same code as long as the button is pressed, the receiver can further improve the reception accuracy by performing a simple "majority vote" for the incoming bit stream: sample a number of incoming bursts and then act on the one that was repeated the most.

Thanks to the active modulation, various message encoding and embedded *error detection and correction* techniques, along with the repetitive sending of the same command, it is almost impossible to end up receiving a wrong code—the selection you make on the infrared remote controller either works correctly or does not work at all, depending on your distance from the receiver and the amount of infrared noise in the environment.

Thanks to all this inbuilt *redundancy*, most devices can handle remote control signals even when the receiver is exposed to direct sunlight, although the functional distance may be reduced. Any background infrared radiation is totally random, lacking modulation and structure, and hence does not match the receiver's expectations. Only if this *infrared noise* is strong enough to totally saturate the receiving circuitry, will it be able to block the receiver, but even then, triggering a false command is extremely unlikely.

The added convenience of being able to switch channels on your television without lifting yourself from the sofa became a necessity around the time when more and more television channels started appearing, along with the ever-increasing bombardment of television advertising.

The person every coach potato has to be thankful to is Eugene Polley: he created the first, simple, light-based remote-control system. Polley was an engineer working for the television set manufacturer *Zenith*, and he got a $1,000 bonus for his invention. To put this in context, this was worth about two television sets at the time, but has an inflation-adjusted value of almost $10,000 today.

His first version, *Flash-Matic* that went on sale in 1955, did not use infrared light: it was a simple combination of a flashlight with a sharp, optically focused beam and four light-sensitive sensors at each corner of the television. You could turn the television on or off, mute audio, and change channels up or down, simply by pointing the light at the corresponding corners of the television set.

As the channel changers of the time were mechanical, this system needed a small motor that rotated the channel selector of the television, mimicking a manual channel switch.

*Flash-Matic* was offered as an add-on for *Zenith's* television receiver, and it wasn't cheap, adding about 20% to the cost of the receiver. Hence, being a couch potato had a high price in the 1950s, which may have contributed to the fact that being overweight is much more common these days than it was a bit over half a century ago.

*Zenith* had already pioneered the remote control of its televisions in 1950 with a wired remote, which was very descriptively called *Lazy Bones*, but customers complained about tripping on the connecting wire in their living rooms.

As *Flash-Matic* was based on simple, non-modulated visible light, direct sunlight posed a problem, sometimes flipping the channels or adjusting the audio on its own. To improve the reliability, the next version that was introduced a year later used *ultrasound* instead of light.

Ultrasound is a sound with a frequency that is beyond the human hearing range, which tops at around 20,000 Hz.

The new remote controller version had multiple metallic rods that were hit by a tiny hammer as the buttons were pressed. These rods were tuned to vibrate on different ultrasound frequencies when hit, and the receiver then turned these high-frequency pings into individual commands for the television.

Although the users did not hear the actual ultrasound ping of a command, the physical tap of the hammer mechanism made an audible clicking sound. This gave rise to the nickname "clicker" for remote controllers for years to come.

On the television side, this new version, boldly named the *Zenith Space Command*, required electronic circuitry consisting of six vacuum tubes, which made it more complex than many of the radios of the era.

The actual remote controller, though, was a purely mechanical device and did not require batteries, as *Zenith*'s management was worried that if the batteries ran out, the users would think that there was something wrong with the television set.

With the earlier *Flash-Matic*, dead batteries would be obvious due to the missing light beam, but with ultrasound, there would not be any obvious indication for exhausted batteries.

The *Space Command* system was created by *Zenith* physicist Robert Adler, and it turned out to be a very reliable way of remotely controlling the television set. *Zenith* kept on using the same approach all the way to the 1970s, and despite the fact that Polley had come out first with the idea of wireless remote control, Adler eventually became known as the "Father of the Remote Control".

This caused some bitterness for Polley, but in 1997, both Polley and Adler were given an Emmy award by USA's *National Academy of Television Arts and Sciences*, for the "Pioneering Development of Wireless Remote Controls for Consumer Television".

Eugene Polley had a somewhat mixed view of his invention—in an interview he gave later in his life to *Palm Beach Post*, he said:

> Everything has to be done remotely now or forget it. Nobody wants to get off their fat and flabby to control these electronic devices.

With the arrival of cheap transistor technology, the mechanical remote controller was superseded by a version that still relied on ultrasound, but generated the commands electronically.

The next step in the evolution of remote controllers was driven by the increasing functionality of television sets: in the 1970s, the *BBC* experimented with their *teletext* service, *Ceefax*, which allowed the continuous sending of data for simple text pages of information during the *vertical blanking interval* between the frames of the television signal. This was another backwards-compatible addition along with the same kind of terms as the adding of the *color burst*, as explained in Chapter 5: *Mesmerized by the Moving Image*.

This new feature brought in the need to perform more complex remote-control operations, like three-digit page number selection and display mode switching, and the total number of these new commands was beyond what could be reliably implemented with ultrasound.

The first *Ceefax*-capable receivers initially went back to the wired remote approach, which had all the known limitations of the very first remote-control systems.

To remedy this, a task force was set up to provide a solution to this problem.

Together with the *ITT Corporation*, *BBC* engineers prototyped a version based on pulsed infrared bursts, and the result of this experiment was the first standardized infrared control protocol, *ITT Protocol*.

This ended up being used widely among many manufacturers, but it sometimes got false triggers if other infrared transmitting devices were around. This was due to the fact that it did not yet use the concept of a carrier frequency, thus lacking one crucial layer of effective signal verification. Later encoding systems eventually fixed this issue.

Infrared transmission is not only applicable for remote control use, as the infrared beam can be modulated with whatever signal you desire. As an example, there are wireless headphones that use an infrared link between a transmitter and the headphones, although Bluetooth is increasingly making this segment obsolete.

The benefit of using infrared instead of Bluetooth, though, is that as this transmission is unidirectional, one transmitter can feed as many headphones simultaneously as can be crammed in the transmitter range. But as was discussed in Chapter 11: *Home Sweet Home*, the latest Bluetooth specification also now supports two simultaneous devices, like headsets, to be used in parallel, so this advantage of infrared audio is also slowly but surely being phased away.

Any kind of constant infrared stream, like the aforementioned infrared based audio distribution, is bound to reduce the operational distance of any other infrared based systems in the same space by adding a high level of background noise, in the same way as excessive sunlight affects the range of infrared remote controllers.

For short-range data transmissions between devices, an infrared protocol called *IrDA* was introduced in 1993. This was a joint product of about 50 companies, which together formed the *Infrared Data Association (IrDA)*, which gave the name to the actual protocol.

With IrDA, it was possible to transmit data between various devices like laptops, cameras and printers. After the introduction of Bluetooth and Wi-Fi, IrDA has practically vanished from use, as all new portable mass-market devices have replaced the former IrDA ports with inbuilt radio solutions, but a majority of the early generation smartphones came equipped with an IrDA port, which sometimes could be used as a remote controller via a special application.

Despite the demise of IrDA, light-based communication is far from dead:

The latest addition is *Light Fidelity (Li-Fi)*, which offers very high-speed, bi-directional data transmission based on visible light.

Li-Fi became possible when conventional light bulbs started to be replaced by LED-based devices, which, being *semiconductors*, can be modulated with very high frequencies. The "flicker", caused by any Li-Fi modulation solution, is millions of times faster than the human eye can perceive, so the users will not notice any negative effects in the overall lightning that also carries Li-Fi data. Currently this technology is only at a very early commercial level, but it has been demonstrated with data transmission speeds of over 200 Gbps.

The first nascent commercial Li-Fi solution provider, *pureLiFi Ltd*, offers a product that provides a data speed of 40 Mbps, which is on par with low-end Wi-Fi equipment.

One strong benefit of Li-Fi is the fact that like infrared, light does not pass through walls, so you will not get interference from your neighbors, as is the norm with Wi-Fi. This naturally also limits the range to confined spaces only, although the fact that light does reflect from walls can make it possible to have Li-Fi reach devices that reside "around the corner".

Li-Fi was pioneered by Harald Haas when he was a professor at the University of Edinburgh. He and his team worked on the groundbreaking *D-Light* project that started in 2010 at Edinburgh's *Institute for Digital Communications*, and after a successful proof-of-concept, Haas co-founded *VLC Ltd* in 2012. The company later changed its name to *pureLiFi Ltd*, and Haas is currently working as its Chief Scientific Officer.

The company is still very much in the early phases, but it has been able to attract considerable venture capital: in 2016, it got a 10-million-dollar fund injection to support the commercialization of this technology.

This *Visible Light Communications (VLC)* technology has big growth potential, providing a potential alternative to the congestion that is becoming increasingly noticeable with the spread of Wi-Fi networks.

The key issue, as with almost any new technology these days, is to agree on a common standard, and this work is currently being done in the *Li-Fi Consortium*. Having a common standard will create the required safe environment for manufacturers to start producing the necessary electronic components that can be integrated into mass-market devices. To enable a new light-based communications paradigm, standardized and compatible Li-Fi components must become as cheap and common as Wi-Fi radios are these days, and the mass-market adoption will follow.

Exactly the same development happened with Wi-Fi, Bluetooth and GPS solutions, and thanks to the high data speeds that Li-Fi could provide, the expectation is that visible light will be the next area pushing the limits of wireless communications.

As was briefly discussed in conjunction with barcodes in Chapter 12: *"Please Identify Yourself"*, light has yet another trick up on its sleeve:

By means of *stimulated emission*, a theory based on Albert Einstein's work in 1917 and first achieved in practice in 1947, a very *coherent*, or unidirectional, beam of light can be created.

This effect is more commonly known as *light amplification by stimulated emission of radiation (laser)*, and it has become one of the cornerstones of modern communications: high-speed *fiber-optic data cables* that crisscross the Earth use laser light as the data carrier.

Thanks to the high frequency of the light, which in its turn supports very high modulation bandwidth, the data transmission speeds are hundreds or even thousands of times faster than can be achieved by copper cables.

Lasers have become so cheap and ubiquitous that for a couple of dollars you can purchase a laser pointer that can be used as a "virtual pointer stick" for presentations. The same handy device can also be used to drive your cat into an endless frenzy while she's chasing that bouncing little red dot.

Unlike natural light sources that randomly spread their radiation to all directions, laser light keeps its tight form across enormous distances. No lens system can compete with the uniformity of a laser beam: it is so tight that if it is shone on the Moon, it will spread to a circle of less than ten kilometers.

As the distance to the Moon is about 400,000 kilometers, this is a very impressive result.

The application of lasers in communication is not only limited to barcode readers and fiber-optic data cables: from the wireless connectivity point of view, laser light has also been successfully utilized for communications between various satellites and the Earth, both residing in Earth orbit and further out in the Solar System.

The first demonstration of laser light-based space communications was done by the *European Space Agency (ESA)* between a ground station and the geostationary *Artemis* satellite in 2001. During these tests, *ESA* achieved gigabit transmission speeds for ground-to-space communications, whereas experiments by the *National Aeronautics and Space Administration (NASA)* have succeeded in achieving space-to-ground speeds of around 400 Mbps.

NASA also holds the record for the longest two-way laser communication link with their *Messenger* space probe, occurring at a distance of 24 million kilometers. This is almost 700 times further out than the geostationary orbit, and about one tenth of the average distance between Earth and Mars.

With potential savings both in terms of size and energy consumption, together with the higher available data transmission speeds, *ESA* and *NASA* are actively looking into the use of lasers as a replacement for or a complement to radio communications in their future interplanetary probes. It's highly likely that whenever mankind makes it to Mars for longer periods of time, the back-bone data link between Earth and Mars could be based on lasers.

Don't expect to interactively browse websites on Earth from Mars though: the time it takes for a signal to cross the distance between Earth and Mars varies between 4 and 24 minutes, depending on the relative position of the planets.

The speed of light may feel enormously fast, but the Universe is way larger still, even within the limits of our own Solar System. As an example of the vastness of our Universe, in 1973, the huge radio telescope of the *Arecibo Observatory* in Puerto Rico was used to send a simple "hello" message towards the globular star cluster *M13*.

This *Arecibo Message* consisted of 210 bytes of binary information, carrying, amongst others, highly simplified descriptions of the double helix structure of the DNA, the human form and dimensions, and the structure of our Solar System.

The transmission of this message, which is the longest distance wireless com- munications attempt ever, will reach its destination in about 25,000 years from now. Whether anybody will ever receive it at *M13* is another issue altogether, and any reply will naturally take another 25,000 years to complete.

As a comparison, we humans were still living in the final period of the *Stone Age* 25,000 years ago.

Similarly, our furthest space probe so far, *Voyager 1*, which was launched in 1977 and is currently traveling with a speed of 17 kilometers per second, will reach our nearest neighboring star, *Proxima Centauri*, in about 73,600 years.

The most amazing thing about *Voyager 1* is that after 40 years in space, it is still functioning, sending measurements back to Earth via its puny 22-watt transmitter, as well as receiving commands from Earth through its 3.7-meter microwave antenna. Even though its signals travel at the speed of light, the time it takes for Voyager's transmissions to reach the Earth is almost 20 hours.

Yet, thanks to our accumulated experience in utilizing the electromagnetic waves, along with the mathematics based on James Clerk Maxwell's theories, it was possible to calculate the necessary antenna sizes and transmission powers in advance, so that we would still be able to receive signals from such a distance almost half a century later.

This is another example of how hard science removes guesswork from the equation.

# Epilogue and Acknowledgements

First and foremost, I want to thank my parents.

Almost half a century ago, they brought me a book about basic electronics circuits from the local library, with a comment "you might be interested in this". After building the first, simple flip-flop circuit that blinked a tiny incandescent lamp, I was hooked, and that book sparked my lifelong interest in everything electronic.

What followed was a string of implementations, from illegal FM transmitters with three-kilometer range to detonation timers for dynamite-based "experiments" with enthusiastic friends: being a teenager in a small, rural village in the back country of Finland had some interesting side benefits at the time, and surprisingly, everyone got away alive, with eyes and fingers intact...

Naturally this is a credit to the public library system in Finland, which has been and still remains world class, nowadays even maybe a bit over the top, as it has been stretched way beyond the original idea of free access to books, newspapers and magazines. But the bottom line is that the variety of material available in the tiny, municipal library of Rasila, Ruokolahti, already during my youth in the 1970s, was simply amazing.

The library even had a special bus that took a selection of books to people in the more remote villages, and if you did not find the book you were looking for, you could order it to arrive on the following week's bus round.

A lot of discussion lately has been centered around the comparative quality of education systems around the world, and on the attacks on science that are happening due to the growing push of faith-based "facts" into the curriculum. This kind of activity has been prominent especially in parts of the United States, often backed by huge amounts of money from highly religious donors, many of which had actually made their fortunes by applying hard science to some business opportunities in the past, yet now appear to be actively trying to suppress scientific thinking for the future generations.

© Springer International Publishing AG, part of Springer Nature 2018          165
P. Launiainen, *A Brief History of Everything Wireless*,
https://doi.org/10.1007/978-3-319-78910-1

In stark contrast to this kind of development, the homogeneous, fact-based education that is the norm in the Nordic countries, is currently being slanted towards giving students the necessary base they need for critical thinking later in their lives, in an environment where fact and fiction are becoming increasingly hard to distinguish from each other.

The differences between societies are much deeper than just in education, though: access to good quality libraries that provide ample amounts of both fact and fiction for young minds has certainly in my case been one, very important aspect in the process of shaping my future, and often this goes hand in hand with the appreciation of learning in general.

I see the wireless, global access to the Internet as the next, great equalizer that allows practically anyone anywhere to tap into the existing and ever-expanding information pool at their own pace: the Internet is becoming the new, universal *Library of Alexandria*, for all mankind.

Information is power, and by empowering all citizens of the Earth with access to it, we can achieve great things together. Having the "Internet in your pocket", as enabled by wireless technologies, makes access to this new, ubiquitous virtual library amazingly easy.

As our continued prosperity is so deeply intertwined with the capabilities of these new technologies, it is hard to grasp the growing keenness to willingly throw away all that cold, hard science and start promoting baseless beliefs instead. This kind of development is most apparent in violent, religious groups like *ISIS* and *Boko Haram*, the latter of which loosely translates into "Western education is a sin". These groups despise education and the use of technology, yet, without the results of all the scientific research that has been behind even their most basic weaponry and communications systems, they would be fighting their wars with sticks and stones, and would be wiped out in an instant against a modern force based on advanced technology.

More alarmingly, the new trend of promoting "make believe" over science appears to be a growing rot even in the most advanced societies. It is raising its ugly head on issues like denying global warming and denouncing vaccinations. Both of these can mean fatal consequences to thousands, even millions of people in the longer term.

Another example of incoherent behavior comes from the silly group of people who have decided that the world is flat and spaceflight is just a big lie, propagated by all of the tens of space agencies in the world—according to these *Flat Earthers*, hordes of well-educated, trained professionals waste their lives by meticulously creating fake imagery from "space", to the count of thousands of pictures per day, even creating a non-stop fake video feed from the *International Space Station (ISS)*.

And if you point out the glaring inconsistencies in their Flat Earth view by referencing everyday services that they use, or even the simple fact that ships clearly appear to slide below the horizon as they sail away from their ports, it does not matter: any "scientific proof" is only cherry-picked when it can somehow be shoe-horned to match their world view—everything else is a deception of some kind.

Why is this kind of behavior so prevalent these days?

My assumption is that the vast improvements in our technical capabilities and the growing ease of use of new inventions have separated us too far from the actual roots of all the advancements benefiting our daily lives. The technology behind these new gadgets is superficially incomprehensible, and hence easily ends up lending artificial credibility to alternative explanations that are pure hogwash.

Add to that the fact that consumers are bombarded with movies that make almost everything look capable of breaking the rules of physics, and the boundaries of reality are blurred further: if Spiderman can walk on a wing of a flying jet plane and Wonder Woman does not even get a scratch while charging into relentless machine gun fire, why wouldn't the shape-shifting lizard people who *really* run the world use commercial jets to control us all with those heinous *chemtrails*, too?

The need to be "special" is a common trait for us humans, but unfortunately it often ends up being specialty of the Forrest Gump style, just without any of the kind-heartedness of the character.

Even worse, as peddling these "alternative truths" to a receptive audience can create sources of income, scrupulous players have stepped in en masse to gain financially from this new level of voluntary ignorance: create a wild claim that makes some atrocious event look like part of a conspiracy, and there's apparently enough audience to turn it into a profitable and repeatable enterprise.

In 2018, this has reached a point where even the current President of the United States is re-tweeting provably false material from highly questionable sources, while at the same time calling professional news organizations "fake news". These tweets then get "liked" instantly by millions, as well as parroted as truths by a major television network and a handful of so-called "real news" websites.

As a result, despite our almost unlimited access to knowledge, our collective ability for critical thinking appears to be rapidly diminishing. The ultimate paradox of the newly relinquished power of the Internet is that the people who have abandoned their trust in science are promoting their views by using the very same tools that were made possible by the science they are actively disparaging.

The *Library of Alexandria* was gradually destroyed over a long period of time by mobs commanded by ignorant rulers. Let's not allow an equivalence of that to happen to the Internet in the form of letting ourselves being overwhelmed by intentional falsehoods.

If the approach of basing important decisions on *faith* instead of *fact* becomes the accepted norm, it will eventually cripple the technological competitiveness of the country it is happening in.

In the words of the late Carl Sagan:

> We live in a society exquisitely dependent on science and technology, in which hardly anyone knows anything about science and technology.

With this book, I have tried to add my small share to the understanding of what makes our wireless world tick. It may often look like a cool magic trick, but it is still based on hard, cold physics.

As for digging up material for this book, the amount of information that is currently at everyone's fingertips through the Internet is absolutely mind-boggling: never before has verifying even the most obscure detail been as easy as today.

In my youth, being able to fix problems came only after a lot of experience: you became a *guru* that could solve things because you had spent countless days and nights relentlessly finding the solutions and workarounds to various problems.

Nowadays, the most useful capability is being able to search quickly and effectively, and increasingly, being able to rate the *quality* of the information that you come across.

In terms of most technological and scientific issues, the amount of false data on the Internet is not really significant, at least yet, but the more you shift towards historical and especially political subjects, the more you really need to weigh the *accuracy* of the information source you are using. Even with credible sources, the details sometimes do not match, and the further back in time you go, the larger the discrepancies appear to be.

Successfully evaluating the reputation of your source of information is a mandatory skill on the 21st century, and should be taught in schools all over the world, not only in the Nordic countries: as we have seen in 2016, absolute false-hoods that are spread via the Internet and other media can even turn elections around, with potentially very detrimental results to all of us.

For creating the basic building blocks for the new, virtual *Library of Alexandria*, I have to give credit to *Google*, despite the fact that I am not at all a fan of their corporate approach towards privacy: the way they hook you up into their network of information harvesting the moment you sign up to any of their services is nothing to be proud of.

Although many companies were pioneering the Internet search, it was *Google* that emerged as the best solution for indexing the information of the world, and they at least appear to be actively working on the immense task of ensuring that their search results have credibility.

I just hope that they would not have so obviously given up on their original mantra of *don't be evil*—privacy should be our birthright, not a commodity that is being exorbitantly collected, benefited from and sold off to the highest bidder.

On a personal level, learning to explain complex issues in simple and palatable terms, I remember the thorough discussions about minute technical details that I had in the early days of my computing career with Heimo Kouvo.

Those sessions were great private tutoring from an experienced guru at a time when you just could not find a solution to your problem through a simple web search. Not only did Heimo teach me a great deal, but also he taught me the way to address complex subjects so that they can be understood by a person with much less background in the subject matter at hand. That helped me immensely in many public presentations that I have ended up giving throughout my career.

Closer to the actual effort of putting a book like this together, my first great thank-you goes to my early round editor Grace Ross.

Not having English as a first language unavoidably means plenty of subtle errors and weird sentence structures, and Grace did a great job in cleaning up an early version of this manuscript. Her comments also helped me to improve those parts of the manuscript that were not clear enough for readers with less exposure to the digital world.

Identically, the final copyedit round done by Lyn Imeson was a true eye-opener. Her attention to detail was impeccable.

As for my experience in the wireless domain, my special thank-you goes to my former employer, *Nokia*, and especially to *Nokia's* former CTO, Yrjö Neuvo. He was the guiding light behind the technology that enabled the explosive success of *Nokia*, and he gave me the idea of moving to their Brazilian research institute, *Instituto Nokia de Tecnologia (INdT)*.

I gained great, global insight into the wireless revolution during my time at *Nokia* and *INdT*, and being located in Brazil instead of the corporate headquarters in Finland also gave me a great, combined outsider/insider view into the decline of *Nokia's* groundbreaking handset division.

I'm very pleased that despite the missteps that doomed their handset division, *Nokia* has been able to continue as the leading wireless network infrastructure provider. Their research unit also seems to be actively looking into new, interesting areas, this time hopefully trying to exploit them better than during the handset days: so many of the great R&D projects inside *Nokia* never saw the light of day.

When this book started getting a more solid structure, the first readthrough of my rough manuscript was done by my friend and former colleague André Erthal.

André eats scientific literature for breakfast, and has the right kind of endless curiosity towards all new things in technology, as well as life in general, and his initial feedback helped me fix the direction of the story.

Early feedback came also from my lifelong friend and fellow computer guru, Yrjö Toiviainen. His "kyllä tästä vielä kirja tulee" (yeah, it will be a book one day) was the kind of traditional low-key Finnish thumbs up that helped me kick the can forward.

Some of the technical aspects regarding the cellular evolution were reviewed by my former colleagues at *Nokia* who prefer to remain anonymous: this does not stop me from expressing my greatest appreciation for their valuable insight. You know who you are...

It took about two years to get this story together, and making a list of every book, article, newsreel, video and website I came across during this time would be impractical. My idea was to make a collection of interesting and entertaining stories, not a scientific study.

To learn more about some of the issues touched in this book, here are some other publications that I would suggest as excellent further reading:

My personal favorite from all the persons mentioned in this book is by far Nikola Tesla.

All in all, Tesla remains one of the most versatile inventors of the 20th century, and his work in wireless technologies is just a small part of his lifetime achievements.

You can entertain yourself for days with the conspiracy stories around Tesla's inventions, or you can get a balanced and comprehensive picture of Tesla's life by reading *Tesla: Man out of Time* by Margaret Cheney.

Encryption has become a fundamental requirement for modern communications, and the use of encryption in human history started much earlier than most of us think. This fascinating part of our history is described in *The Code Book: The Science of Secrecy from Ancient Egypt to Quantum Cryptography* by Simon Singh.

Last but not least, going back to the comments I made above about the alarming increase in nonsensical pseudo-science, I don't know of a better book that approaches this issue than *The Demon-Haunted World: Science as a Candle in the Dark* by Carl Sagan.

Denying or artificially diminishing the findings of science for whatever reason is ultimately detrimental to the goal that should be common to us all: making this planet a better place for all of us to live and learn.

I hope that in this book I have done my tiny bit in trying to give insight into these developments we are so utterly dependent on today. They may look simple on the outside, but this is just an illusion, created by the thousands and thousands of great minds that have spent sometimes their entire lives in improving the underlying technologies.

Therefore, my final thanks go to all of the named and unnamed heroes that have made all this possible, making this a better world for all of us.

# TechTalk
# Sparks and Waves

From a technology point of view, the very first radios were crude electro-mechanical devices, utilizing coils, transformers and the so-called *spark-gap* approach to generate radio waves.

Spark-gap is basically just a method of generating radio frequency noise through an actual stream of high-voltage sparks, and to make matters worse, the resulting generated frequency is not very precise.

Therefore, the transmissions would "bleed" to adjoining channels, making it difficult to use a receiver tuned on a different channel in the vicinity of a spark-gap transmitter. Two transmitters on adjoining channels would easily cause considerable interference to each other, and although adding more transmission power would expand the range, it would also expand the radius of the potential interference.

During the early years, this nasty side-effect was unfortunately often used to deliberately interfere with competitors' transmissions in various contests and demonstrations between rival equipment manufacturers, unnecessarily tarnishing the perceived usefulness of radio.

There was no *modulation* for the generated signal: the transmitter was turned at full power and back to zero power via the *telegraph key*, so effectively the telegraph key was used as a nimble on-off switch for the spark-gap transmitter, producing high-frequency pulses that were interpreted as *Morse code*.

In Morse code, each letter is represented by a set of short and long pulses, *dots* and *dashes*, and these combinations were optimized according to the statistical representation of letters in the English language. In this way, the overall time needed to transmit English text as electric pulses was minimized. For example, the most common letter is *E*, and it is presented as a single short pulse (dot) in Morse code, letter *I* got two short pulses (dots), letter *A* one short (dot) and one long (dash), and so on.

Naturally, text in German or Swahili does not share the same statistical distribution of letters, and hence would not be optimized as Morse code, but the same

© Springer International Publishing AG, part of Springer Nature 2018
P. Launiainen, *A Brief History of Everything Wireless*,
https://doi.org/10.1007/978-3-319-78910-1

letter coding is used globally anyway, with new, lengthy codes added for the local special letters.

This pulse-based communications method was already in use in the telegraph lines crisscrossing the world, so it was natural to use the same approach with radios as well. Finding persons that were fluent in Morse code was easy, thanks to the pool of existing telegraph operators.

As systems were purely electro-mechanical, it was difficult to reduce the amount of interference that was generated by the transmitter. Considerable improvement was only possible when vacuum tube technology become available years later.

Things were not much better on the receiving side:

In the early designs, a component called a *coherer* was used to detect the radio signal. This consisted of a tube with two electrodes, filled with metal filings, the conductivity of which changed in the presence of electromagnetic waves that entered the coherer via the connected antenna.

In this way, the presence of a *carrier wave*, the unmodulated high-frequency transmission, was detected, and the resulting change in electrical conductivity could be used to drive a ticker tape mechanism or feed a sound to headphones.

In order to reset the metal filings inside the coherer for another pulse, a mechanical tap on the coherer was needed. This "knock" returned the filings back to their low-conductance state. Therefore, the coherer was very susceptible to both mechanical and electric disturbances, and a lot of effort was put into the design of the necessary *decoherer* assemblies that were needed to reset the coherer after each pulse.

The first coherer was invented in 1890 by a French physicist Edouard Branly, and the inner workings of it and its successors were a kind of black art of the era: progress was made mainly through an endless loop of tweaking and retesting by the radio pioneers of the time.

In hindsight, having all this hodge-podge circuitry providing communications over thousands of kilometers was another amazing example of how human ingenuity relentlessly pushes the limits of available technology, however primitive it happens to be. Even today, the physics that make the coherer work are not fully understood, yet it did not stop the inventors utilizing it as a key component of a radio receiver, constantly testing new designs in order to improve the selectivity of the receivers.

The next major step was to resolve the problem of interference that was caused by the wide frequency spectrum of the carrier wave generated by the spark-gap transmitter technology. To cure this, the research focus moved to developing highly tuned, *continuous wave* transmissions. These systems would be able to generate a *sinusoidal wave*, a pure signal of a fixed frequency, greatly reducing the broad-spectrum radio noise created by the spark-gap.

Again, the earliest continuous wave attempts were electro-mechanical, using traditional alternating current generators, but with higher rotating speeds and denser coil structures that were able to generate radio frequency oscillations. The most notable of these systems were the *Alexanderson Alternators*, which were powerful enough to enable cross-Atlantic transmissions.

The generational breakthrough away from electro-mechanical systems came after the introduction of the *triode*, which was invented in 1907.

A triode, as the name implies, is a *vacuum tube* that has three embedded metallic terminals sealed in glass tube that has all air removed from it. Current flows through two of the terminals, the *cathode* and the *anode*, and a controlling voltage on the third one, the *grid*, can be used to increase or reduce this main current flow. The fundamental, groundbreaking feature of a triode is that a small variation of the control current can cause a much higher change in the flow of current from cathode to anode, and hence, for the first time, it was possible to amplify the weak signal that was received through the antenna.

The downside of a triode is the fact that in order to maintain a flow of current, you had to provide internal heating of the cathode through a special, glowing filament. This filament, just like in conventional light bulbs, has a limited life span, and when it breaks, the triode is no longer functional. For commercially produced triodes, the expected life span varied between 2,000 and 10,000 hours. Hence, in the worst case you had to replace them after three months of continuous use.

Heating the filament required considerable extra energy that was in no way participating in the current flow of the actual circuitry. This was problematic for any battery-operated devices. Finally, the glass tube that was used to contain the vacuum in a triode was very fragile against physical shocks.

Despite these shortcomings, the invention of the triode revolutionized radio technology by moving it into the *solid-state electronics* era: the equipment no longer needed bulky and complex moving parts, thus becoming more robust, smaller, and superior to the first-generation devices in terms of transmission signal quality and receiver selectivity. Large-scale production of vacuum tubes brought their price down rapidly. All in all, this fundamental change was behind the *broadcast revolution*, as discussed in Chapter 4: ***The Golden Age of Wireless***.

The second revolution in solid-state electronics started in 1947, with the invention of the *transistor*.

The original transistor worked on the same principle as a triode, offering a controlling electrode, the *base*, which could be used to govern a much stronger current flow between the *emitter* and the *collector*. But as transistors are based on materials called *semiconductors*, they do not require a glass tube with an internal vacuum, or a special, power-hungry heating filament, and are therefore much smaller and considerably more energy-efficient than triodes. Without a filament, the lifespan of transistors is in all practical terms unlimited, as long as their operating conditions stay within specifications.

The most common material to provide the semiconducting material for transistors is *silicon*, which is abundant and cheap, and unlike the case with vacuum tubes, the manufacturing of transistors can be easily scaled up to very large-scale production: as a result, the cost of an individual general-purpose transistor drops to a level of less than one cent when purchased in bulk.

The most fundamental benefit that arose from the improving manufacturing processes for transistors was the possibility to create complete *integrated circuits*, consisting first of thousands, and currently billions of interconnected transistors.

These *microchips* are behind our current, computer-driven world, and they come in all shapes, sizes and functionalities: if you go to the website of the microchip reseller company *Digikey,Inc.* and search for "integrated circuits", you get over 600,000 hits. Many of them perform just one specific electronic task, but one important subset of microchips, *microprocessors*, has made it possible to add generic computational functionality to almost every device today.

Transistor technology has gone through several generational improvements over the years, resulting in a stream of constantly cheaper and more energy-efficient components. Thanks to large-volume production methods, the price of the simplest microprocessors, consisting of hundreds of thousands of internal transistors, is below one dollar.

By altering the internal organization and wiring of transistors inside a microchip it is possible to provide almost limitless variation of highly specific functionality in a very small space. Good examples of this kind of specially designed component are the *graphics processors* that are able to provide us with an immersive, real-world-mimicking 3D-experience in the latest *Virtual Reality (VR)* systems.

At the other end of the scale are *microcontroller* chips, which combine the logic functionality of microprocessors with suitable interfacing circuitry that makes it possible to create versatile devices with minimum external components. These can be programmed to perform whatever functionality the user desires.

With microchips, the human imagination is the final limit, and our current, transistor-based technology appears to have lots of room for feature expansion in the coming decades.

Any kind of electronic component that can amplify signals can also be made to oscillate on a highly precise frequency through suitable feedback circuitry. Therefore, the introduction of triodes was a death knell for all those bulky electro-mechanical components that had earlier been used for radio frequency signal generation.

It was not only possible to produce pure, continuous sinusoidal waves, but the triode also enabled the generation of much higher frequencies than had been possible with its electro-mechanical predecessors. Hence, more channels opened up for communications.

Pure sinusoidal waves, as well as precise control of the transmission frequency, resulted in a huge reduction of interference between channels. These were both superior features compared with the earlier spark-gap transmitters, and this development only got more energy-efficient and reliable with the introduction of transistors.

On the receiving side, a major breakthrough came in the form of *super-heterodyne* technology, an implementation by Edwin Armstrong in 1918, which was based on the theoretical *Heterodyne Principle*, patented by Reginald Fessenden thirteen years earlier.

Directly amplifying the high-frequency signals that are received by an antenna is difficult because the higher the frequency, the lower the gain that can be achieved, even by active electronic components.

In a super-heterodyne receiver, the weak high-frequency signal from the antenna is mixed with a stable, low-power frequency from a local oscillator, and this mix then creates an intermediate, much lower frequency that can be amplified easily. This ingenious way of amplifying the intermediate, low-frequency signal instead of the directly received high-frequency signal results in a superior reception sensitivity, which in turn means that radio stations that reside much farther away can now be received, without the need of increasing their transmission power.

Along with the possibility of selecting the frequency with precision, the new transmitters enabled the use of *Amplitude Modulation (AM)*: instead of having the transmitter just on or off, as controlled by the telegraph key, it was possible to adjust the transmission power dynamically between minimum and maximum. Thus, by using an audio signal as the driver for the Amplitude Modulated transmitter, the spoken word could be used for communications instead of the relatively slow Morse code.

Although voice soon took over the majority of radio communications, Morse code remained relevant all the way to 1999, when it ceased to be used as an international standard for long-distance maritime communications.

The remaining notable uses for Morse code are *Amateur Radio* communications, and the identification of *VHF Omni Range (VOR)* radio navigation aids in aviation, which were discussed in Chapter 6: **Highways in the Sky**.

Modern aviation equipment automatically turns the received three-letter Morse code into text on the navigation equipment's screen of the airplane, relieving the pilot from doing the station's Morse code identifier verification by ear, but the corresponding dots and dashes for VOR station identifiers are still printed on aviation charts.

Despite this automation, the continuing use of Morse code in VOR stations makes it the oldest electronic encoding system still in use in the world.

Amplitude Modulation has an inherent limitation: when the modulating signal level is low, the resulting transmission power is low as well, and the lower the received signal, the more any interference on the same channel reduces the quality of the received signal.

As Amplitude Modulation was also used for analog television transmissions, the interference would be most visible when no picture was transmitted and the television showed just a black screen—a passing motorbike with a poorly shielded ignition circuit would cause random white dots to be visible on the black image.

The logical step to circumvent this problem came by inverting the modulation signal: when the screen was all-black, the transmission was at its strongest level, thus masking any low-level interference that would otherwise be very visible on the black background.

For audio transmissions, a far better way was devised to circumvent this— *Frequency Modulation (FM)*, which was patented in 1933, allows the transmitter to work with continuous maximum power, while the modulating signal is used to cause slight variation to the actual transmitted frequency instead of its amplitude. Hence the received signal is always at maximum level, whatever the momentary modulation level may be, and thus the reception is much less susceptible to spurious interference. Also, the potential transmitting range is at its practical maximum.

Frequency-modulated signal therefore has a center frequency around which the frequency keeps on shifting, and the width of this shift depends on the depth of the modulation. This constant shifting around the center frequency is then detected in the receiver and turned into the original modulating signal.

As explained in TechTalk *There is No Free Lunch*, the wider the frequency spectrum is in the modulating signal, the wider will be the required bandwidth for the transmission channel. Therefore, Frequency Modulation is only practical if the transmission frequency is at least several tens of megahertz (MHz).

But the improved sound quality of FM, in combination with the use of the *Very High Frequency (VHF)* band, was the final improvement that elevated broadcast radio to its current, almost universally available status in the familiar 87.5–108 MHz broadcast band. Only a handful of exceptions exists to this universal standard, most notably Japan with its 76–95 MHz allocation. Some of the former Soviet Bloc countries also used to have a different allocation of 65.9–74 MHz, but most of them have now shifted to the 87.5–108 MHz model.

AM transmissions remain standard for lower frequency bands, on which the signal can traverse long, even intercontinental distances. This is due to the fact that the lower-frequency signals follow the curvature of the Earth as a *groundwave* and also via reflection back from the *ionosphere*, an electrically charged region of the upper atmosphere.

This effect is reduced when higher frequencies are used, and for FM transmissions in normal conditions it has entirely vanished: the transmitted signal moves directly away from the transmitter antenna in a straight line, without reflecting back from the ionosphere and without having a groundwave effect. If the FM receiver is far enough from the FM transmitter so that there is no longer a line-of-sight access to the antenna due to the curvature of the Earth, the reception quality quickly deteriorates, because the signal just passes overhead into space. The relative altitude of the receiver has a bigger effect on the receiving distance than the huge transmitting power that is invariably in use in most broadcasting FM transmitters: if you check your FM radio on a plane flying at 10,000 feet or higher over a densely populated area, your dial is suddenly full of faraway stations—another simple proof that the Earth is not flat.

This direct-line propagation of FM signals is the reason why the antenna towers are usually very tall, or built on hilltops. Depending of the height of the transmission antenna, you can normally only hear FM broadcasts to about 200 kilometer distance, whereas due to the groundwave and ionospheric reflection, AM transmissions can be received from stations that are thousands of kilometers away.

Long-distance high-frequency reception of both FM and television signals can happen only in rare, reflective atmospheric conditions, often related to large high-pressure areas, during which the *troposphere*, the part of Earth's atmosphere where all weather occurs, reflects the signals over longer distances, or when the ionosphere is agitated by higher levels of *solar wind*. Both of these create short-to-medium term reflective situations, during which even the higher frequencies can get mirrored back, reaching receivers that are hundreds or even thousands of kilometers beyond the horizon. As the FM radio network's channel allocation was planned with the line-of-sight limitation in mind, this kind of "radio weather" may result in severe interference in situations where the receiver resides far from the desired transmitter—the reflected signal from another transmitter on the same channel may be strong enough even to mask the other, desired transmission completely.

As this kind of special reflectivity situation varies continuously, and due to the way FM receivers try to lock onto a single signal when two signals appear on the same channel, you may experience a situation where your FM receiver seems to have a mind of its own, flipping continuously between two transmissions.

A very rare occurrence of reflection may also happen if a *meteor* enters the atmosphere in just the right location between you and a faraway transmitter: the *ionized air* caused by the burning meteor may reflect a strong–enough signal that masks a weak FM transmission for just a couple of seconds, and again, it feels like your radio just flipped channels for a brief moment on its own. I personally experienced this kind of *meteor scatter* phenomenon while "hunting" for remote television broadcasts as a kid—I had the television on an unused channel in Finland, hoping to catch some long-range reflections caused by a hot, summery high pressure, and suddenly got a picture-perfect television test picture from Denmark, only for it to vanish after 2–3 seconds, never to be seen again.

As I explain in *Epilogue and Acknowledgements*, growing up in rural Finland made you quite inventive in finding ways to spend your spare time...

For radio transmissions over very long distances, another effective variation of Amplitude Modulation, *Single-sideband Modulation (SSB)*, can be applied. It was patented in 1915 by John Renshaw Carson, but only saw its first commercial use twelve years later as the modulation method for the transatlantic public radiotelephone circuit between New York and London.

SSB has better transmission efficiency than normal AM transmission, thus enabling a longer reception range with equal transmission power.

Amplitude and Frequency Modulation are both well suited to be modulated by analog signals, but when your signal is digital and consists of just a stream of ones and zeros, as discussed in TechTalk *Size Matters*, these original modulation modes are not optimal in terms of both spectral efficiency and susceptibility against interference.

To squeeze the very maximum bandwidth from the available channel, not only the signal frequency and amplitude, but also the phase of the signal is modulated.

The most commonly used digital modulation scheme is called *Orthogonal Frequency Division Multiplexing (OFDM)*. Going into details of OFDM is beyond the scope of this book, so let's just say that it is very efficient in terms of its use of the available spectrum, as well as relatively immune against various common forms of radio interference, like *multipath propagation interference* that is caused by reflections from large objects or mountains.

Hence OFDM is the modulation used in many of the wireless solutions discussed in this book, including Wi-Fi, WiMAX and 4G LTE.

The slice of the *electromagnetic spectrum* that in practice is exploited for radio, the *radio spectrum*, covers frequencies from *Very Low Frequencies* (*VLF*, down to about 3 kHz) to *Extremely High Frequencies* (*EHF*, about 300 GHz).

As the frequency change is a linear progression, the boundaries of the various frequency bands are not absolute: the change from one group to another is gradual, but the commonly mentioned boundaries are as follows:

Very Low Frequency (VLF), starting from 3 kHz
Low Frequency (LF), starting from 30 kHz
Medium Frequency (MF), starting from 300 kHz
High Frequency (HF), starting from 3 MHz
Very High Frequency (VHF), starting from 30 MHz
Ultra High Frequency (UHF), starting from 300 MHz
Super High Frequency (SHF), starting from 3 GHz
Extremely High Frequency (EHF), starting from 30 GHz

If we go below 3 kHz, there is very little room for meaningful modulation, for reasons discussed in TechTalk *There is No Free Lunch*.

The low end of the spectrum has some very special uses, like communicating with submarines, as the VLF frequencies can be received at depths of hundreds of meters, requiring an antenna that can be kilometers long, towed by the submarine.

At the top end of the range, at *Super High Frequencies (SHF)*, we first hit microwaves, which are utilized by a lot of our current technologies, from radar to Wi-Fi to microwave ovens. Thereafter, when passing beyond the upper microwave limit of 300 GHz, we reach the very top portion of the electromagnetic spectrum, where the propagation properties of the electromagnetic waves change radically: the first part is infrared radiation, followed by visible and ultraviolet light, and finally reaching deeply penetrating X-rays and extremely powerful gamma rays.

All in all, this offers us a wide selection of useful frequencies to exploit, with wildly differing properties: lower frequencies can be used for communication between continents and even underwater, whereas high frequencies can pack enormous amount of information, thanks to the wide modulation they can support.

Gamma rays are the result of nuclear reactions, and the most energetic gamma rays can reach us from the other side of the known Universe, created by *supernova* explosions or colliding *black holes*, and they will pass through the Earth more easily than a ray of light passes through clear glass.

The high end of ultraviolet radiation, together with X-rays and gamma rays, present the part of the electromagnetic spectrum that contains *ionizing radiation*, and hence these high-energy rays are harmful to any living tissue. As discussed in Chapter 11: *Home Sweet Home*, microwaves are **not** ionizing radiation, they only generate highly localized heat.

As a summary, the solid-state electronics revolution, with its precisely tuned, high-frequency transmitters and selective super-heterodyne receivers, was the most fundamental breakthrough on the path towards our modern wireless society.

Thanks to this improved technology, legislation was drawn up in 1927 that considerably restricted the use of spark-gap transmitters: due to the advances in technology, the pioneering equipment had suddenly turned into a major source of interference for the receivers of the triode era.

After the invention of the transistor, the fundamentals of radio receivers and transmitters remained the same, but the advances in energy efficiency, reliability and size have kept progress in high gear. And the recent microchip revolution is finally bringing entirely new approaches to this technology, as explained in TechTalk *The Holy Grail*.

Radio waves, by their very nature, are a shared and limited resource, and as they do not adhere to national borders, close collaboration is needed to avoid interference between different users. The *International Telecommunication Union (ITU)* acts as the top-level allocation entity for resolving any potential issues in a global scale, as in the discussion about the allocation of geostationary satellite frequencies in Chapter 7: *Traffic Jam Over the Equator*.

On top of this international collaboration, practically every country has their own regulatory entities that govern the allocation and use of the various frequency bands within the borders of the country, and due to historical reasons, these allocations often differ between geographical areas, practical examples of which were discussed in Chapter 9: *The American Way*.

As the technology developed, the usage of some bands became obsolete and these bands were reallocated, like what happened with the release of some of the former television frequencies, as described in Chapter 5: *Mesmerized by the Moving Image*.

# TechTalk
# Size Matters

All of us are familiar with the concept of analog and digital wristwatches: one has linearly sliding pointers for time, another has distinct numbers that flip in succession.

While these two concepts essentially present exactly the same thing, they offer great examples of what *analog* and *digital* actually mean.

In the analog world, things happen in linear fashion, whereas in the digital world, things happen in fixed steps, and unless we are observing the world around us on an *atomic* or *quantum* level, everything around us seems to be analog: however precisely we measure some natural event, it seems to move from A to B so that every possible point between A and B is touched along the way—the sunlight dies away without any noticeable steps at nightfall and the stars appear linearly brighter and brighter as the sky around them darkens. Even the lightning bolt, when recorded with a high enough frame rate, has a clear start, growth, decay and end.

But when it comes to computers, the situation is very different:

Computers store information internally in *bits*, which only have two states—either zero or one. This is like a light switch: it is either on or off; there is no concept of *half* in computers.

If we were limited to choose something like the volume level for our music system only as *on* or *off*, that would be very annoying indeed, as anyone with teenagers at home has surely experienced. To circumvent this, computers handle any arbitrary data as a collection of bits, grouped into larger units for the convenience of transferring and processing them. What such a collection of bits actually means is totally up to the implementation: together they can present a digitized picture, the event history of a home alarm system, your favorite piece of music, a copy of this book. Anything.

This description pretty much sets the scene for all *digital data*: everything is just a smaller or larger collection of bits, and what these bits represent is totally up to the implementation. And the implementation is just a gentlemen's agreement between

© Springer International Publishing AG, part of Springer Nature 2018                181
P. Launiainen, *A Brief History of Everything Wireless*,
https://doi.org/10.1007/978-3-319-78910-1

man and a machine: we have simply defined that data in a certain format means a certain thing.

In order to bridge our analog world to computers, we need to convert the linearly varying values into a stream of fixed numbers, and as long as we do it with small enough steps and repeat this process quickly enough, our limited senses can't tell the difference.

For example, our standard *Compact Disc Digital Audio (CDDA)* has a resolution of 65,536 steps for the possible volume level on each stereo channel at any moment in time of a piece of music. A value of zero means no sound, whereas the value of 65,535 indicates the maximum theoretical level, which with CDDA is about 96 dB.

Without going into too much detail, let's just note that the 96 dB range, if fully utilized, is good enough for any projected use case for CDDA, considering that a quiet room has an ambient noise level of about 30 dB, and the human pain threshold is around 130–140 dB. Hence if you set up your audio equipment amplification so that even the very lowest sounds levels on your *Compact Disc (CD)* are audible over the ambient noise, the highest levels will just about start hurting your ears.

To convert the original audio into digital audio in CDDA format, the signal on both stereo channels is sliced, or *sampled*, into these 65,536 steps 44,100 times per second. This is called the *sampling frequency*, and the magic number of 44,100 was chosen due to a *Nyquist Theorem*, which states that in order to be able to reproduce the original waveform, a sampling frequency that is at least twice the maximum frequency in the source signal is needed.

And as the top frequency range of human ear is around 20,000 Hz, sampling music with 44,100 samples per second was deemed sufficient by the engineers of *Philips* and *Sony*, the inventors of Compact Disc.

Conveniently, two bytes, or 16 bits, is enough to represent 65,536 distinct values, and hence each sample of a stereophonic signal consumes four bytes. With 44,100 samples per second, storing one second of stereophonic CDDA audio requires therefore 176,400 bytes.

Although the digital audio volume always changes in steps, the steps are so small and happen so quickly that our ears can't tell the steps apart, despite the claims of "golden-eared" audio purists that try to belittle the CDDA quality.

In real life, our analog ears can be fooled with much less quality, as the proliferation of *streaming music services* that rely on *lossy compression algorithms* has proved. For example, both *Spotify* and *Apple Music* rely on lossy compression, meaning that all parts of the sound that are considered non-essential to the perception capabilities of the human ear have been removed. Therefore, if the streamed digital data is compared with the originally sampled signal on a purely numerical level, they have very little in common. But thanks to the *psychoacoustic* limitations of our brains, the more than one hundred million satisfied users of these streaming services could not care less about this very real inherent discrepancy.

These lossy audio compression algorithms first became in wider use through the *MPEG-1 Audio Layer III (MP3)* standard, mainly developed by the Fraunhofer Institute in Germany. Other, more recent standards are *Advanced Audio Coding*

*(AAC)* and *Ogg Vorbis*. The theoretical work in the area of psychoacoustics is nothing new: it dates back all the way to the end of the 19th century.

The process of taking an analog signal, like our voice while we speak into our smartphone or video from a television camera, and turning it into a collection of bits that computers can handle, is called *digitizing*, and the computers around us do this all the time: they can't handle analog signals, so they have no other choice than to employ electronic circuits called *analog-to-digital converters* to do the job.

On the other hand, when computers are required to restore the digitized data back to analog form for human consumption, they use *digital-to-analog converters* to reverse the process.

The additional benefit of swapping analog data into numbers comes from the fact that these sets of numbers can be further processed in a way that reduces the size needed to save and transmit them. As mentioned above, you can greatly reduce the size of digitized information by the use of lossy compression algorithms, but in cases where you have to be able to *exactly* reproduce the original information, several *lossless compression algorithms* can also be used.

If, for example, the digitized data happens to have 5,000 zeros in a row, instead of saving a zero into 5,000 consecutive memory locations, we can define a scheme that allows us to say in shorthand format that "the next 5,000 numbers will be zeros". Depending on the way we define such a lossless compression scheme, the space needed to save these 5,000 numbers can be cut to under 10% of the original size.

We therefore save a lot of storage space or transmission bandwidth, but the payback comes when we access the compressed data, and we have to use computing power to decompress it.

*Digital Signal Processors (DSPs)* can do such compression on the fly, and they are the essential ingredients for all digital communications these days. The control programs for these DSPs are called *codecs*, and they were the behind the technological leap that made it possible to jump from analog first-generation cellular networks to digital second-generation networks.

As discussed in Chapter 8: **The Hockey Stick Years**, during the transition from first-generation to second-generation cellular networks, we were able to squeeze three digital voice channels into the same amount of precious radio spectrum that formerly could handle just one analog one, thanks to the real-time conversion capability of codecs. Therefore, with all other aspects being the same, three times more users could be handled in the same amount of bandwidth.

There are many other cases where digitalization and subsequent compression create huge savings in terms of required memory space, but they are beyond the scope of this book.

But let's go back to the size issue:

As mentioned, a single bit is the smallest piece of information a computer handles, but for practical purposes, computer memory is designed to handle data in bigger chunks, most notably as *bytes*, which consist of eight bits.

As each of the eight bits in a byte can be either one or zero, a little experimentation with paper and pen would reveal that one byte can have 256 distinct values as the individual bits switch to ones and zeros in all possible combinations over the eight available bits. 256 variations are enough to cover the *American Standard Code for Information Interchange (ASCII)* encoding of all letters, numbers and special characters in text written in English, actually leaving 128 possible bit combinations in a byte still unused.

The problem comes with special letters in other alphabets:

Thanks to all the different ways us humans have invented for representing written text in different languages around the world, the number of possible combinations does not fit in the available extra 128 variations inside a single byte.

Hence, the Internet as we know it most commonly uses a text encoding standard called *UTF-8* for the presentation of the pages that we browse daily. Just to prove that real engineers named this format, UTF-8 is shorthand for *Universal Coded Character Set Transformation Format–8-bit*, and it is a dynamic-length character representation: the lowest 128 variations match the original ASCII table, but on top of that, a single character can require anything from one to five bytes to be identified.

For example, the euro currency character, "€", requires three bytes in UTF-8, with values 226, 130 and 172, whereas the letter "a" needs only one byte with a value of 97, matching its value in the ASCII standard.

Therefore, in order to represent letters "a€" in the UTF-8 format, we need space for four bytes, the numerical values of which are 97, 226, 130 and 172.

This combination of consecutive byte values only means "a€" when the content is handled as UTF-8 text, so the computer has to know somehow that we are dealing with text, and that the text should be interpreted in UTF-8 format.

Like this example, any data that a computer is using is eventually broken into chunks of bytes, and when these chunks are stored, there is always some agreed-upon mechanism that tells the computer what the current bunch of bytes actually represent: a text file, a picture, a video, a spreadsheet etc. The way the computer is made aware of this is totally implementation-dependent, and can be as simple as a certain suffix in the file name, or a set of specific bytes in the beginning of the file itself.

Therefore, in essence, all data in computers is a bunch of zeros and ones, usually grouped into sets of eight bits stored in a byte, which is the smallest individual unit that the computer accesses internally.

This book, with all the formatting and other information needed to represent a text file for the *LibreOffice* program that was mainly used to write the original manuscript, fits in about 900,000 bytes.

Playing with such big numbers is hard, so we use suitable prefixes to ease the number game:

900,000 bytes can be presented as 900 kilobytes, "kilo" meaning a thousand of something, as in *kilogram*, which is thousand *grams*.

900 kilobytes is usually written as 900 kB.

The next common multiplier is the *megabyte (MB)*, which is one million bytes. Hence this book with its 900,000 bytes can be said to contain 0.9 MB of data.

When you buy a common *USB Memory Card*, or a *Micro SD card*, which are both pretty much the same thing in terms of their internal storage technology, they are currently sized in *gigabytes*.

One gigabyte is one thousand megabytes, or if represented in short form, 1 GB = 1,000 MB.

For example, at the time of writing this, you could buy a 16 GB memory card for under five dollars, and fit roughly 16,000 copies of this book on it.

So, 16 GB looks big enough to last a lifetime?

Unfortunately, no.

Text is very compact in terms of its memory requirements, but things look different with other types of data: snap a picture with your reasonably good smartphone, and depending on the complexity of the picture, you need 2–5 MB to represent just one image. The actual size varies due to the lossy compression that is used to minimize the amount of space required, and the level of compression that can be applied depends on the structure of the picture.

Hence you can save about 4,000 pictures on a 16 GB memory card.

That still sounds like a big capacity for an average user, but a quick count in my own image repository shows that I have about 16,000 pictures archived so far.

All of them would still fit on four 16 GB devices, or a single 64 GB one, which is a size that is also readily available today for roughly 20 dollars.

But if you turn on the video feature of your smartphone, the requirements for data storage jump again: for one minute of *High Definition (HD)* quality video, you need roughly 100 MB of memory.

Suddenly your 16 GB memory card can only hold about 15 minutes of video.

Not quite enough for a lifetime of video memories.

Add to that new, ultra-high resolution *4K* video and whatever 360-degree *Virtual Reality (VR)* video formats will be used in the future, and it is obvious that storage requirements for data never end.

Luckily enough, the price of memory is going down all the time, while the available capacity is going up: it was not so long ago that a 64 GB memory card cost about 100 dollars, and if you go back fifteen years, having 64 GB on a memory card was pure Science Fiction.

For larger capacities, you need to shift from memory cards to other types of memory, like *hard drives*, and at the time of writing, it is possible to buy hard drives with 10 terabyte (TB) capacity, where 1 TB = 1,000 GB.

And both the available size of these various memory devices, along with the new ways to fill up that space keeps on progressing. Relentlessly.

When it comes to the memory that is in use by the computer or smartphone to accommodate the actual processing of data by the applications used, we are dealing with smaller numbers.

Although the requirements for smooth operation tend to go up continuously as our expectations for our portable devices grow, we can currently manage by having between 1 and 16 gigabytes of device memory to run our applications.

More is always better here, but the reason why we can't have as much as we like is because this memory is different and more expensive than what we discussed earlier.

The type of memory in devices like USB Memory Cards and the so-called *Flash Memory* in smartphones, as well as the *Magnetic Memory* in traditional hard drives, is *non-volatile*: whatever was written in that kind of memory is retained when power is turned off.

In contrast, the memory used by the operating system of a smartphone while running applications is *volatile*, meaning that when you turn the smartphone off, whatever was in the volatile operating memory is wiped out. The benefit of being volatile is that both reading and writing access of this kind of memory is very fast compared with most non-volatile memory access times, allowing the processor to run applications with maximum speed.

The slower, non-volatile memory is only accessed when data or applications need to be loaded or stored: all other processing is done with maximum speed inside the volatile memory.

Some types of non-volatile memory also have limitations on how many times a single storage position can be written over before it wears off and fails, but for most practical uses, this number is so large that it can be ignored.

Last but not least, a practical note: whatever capacity or type of long-term storage you have, **always have at least one backup**, more preferably two, and to be on the safe side in case of fire or theft, **do not keep your backups in the same place as your computer.**

Swap one copy every week with another memory card that you keep in your work drawer. Or give it to a friend, or save your precious data to a *cloud backup service*.

Things break, are stolen or get accidentally deleted, usually just when you least expect them to. And remember, just having a *mirrored copy* somewhere in the cloud is not enough: if you accidentally delete your local copy, the deletion may be mirrored in the cloud the moment you synchronize your device again. There goes your copy...

Save your work and save it often—computers have a tendency of breaking one way or another, just when you need them most.

Also make multiple versions of your work: don't just save with the same name over and over again. There is no limit to version numbers you can use, and this approach allows you to go "back in time" if you inadvertently break something along the way: I ended up saving almost 400 intermediate versions while writing this book, and due to an error I made during one late-night session, one of them was worth all the extra hassle of versioning.

The discussion about mirroring and copying our digital data brings us neatly back to the subject of wireless communications: for *uploading* or *downloading* all these books and pictures and videos and whatnot across any data transmission connection, we use *bits per second (bps)* to present the available *data transmission speed*.

Without taking into account any overheads needed for ensuring successful data transmission, you can deduct from the numbers above that sending one *megabyte* of data over a channel that provides 1 megabit per second speed takes approximately eight seconds, as every byte has eight bits to send.

With all of the overheads needed for data encoding, splitting the data into individual packets and various *error detection and correction* schemes, a multiplier of 10 is quite OK to use for general calculations.

In an *error detection and correction* scheme, additional information is embedded to the transmitted signal, which can be used to mathematically verify that the data was received correctly, and in some cases even restore the original signal if the detected error was not large enough.

The speed that is available for wireless communications depends on various aspects like the frequency band in use, available bandwidth per channel, number of parallel users on the same channel, modulation type and distance between transmitter and receiver. All these cause physical limitations that affect the potential maximum speed of a transmission channel.

As an extreme example, when *NASA's New Horizons* spacecraft passed by Pluto in 2015, it gathered over 5 GB of images and other data on Pluto and its companion Charon, storing it all in its non-volatile on-board memory.

All this data collection and picture grabbing during the fly-by happened automatically in just a matter of hours, but as *New Horizons* was only able to transmit the collected data back to Earth with an average speed of 2,000 bps, it took almost 16 months until the last bit of data was safely received by *NASA*.

Had something gone wrong with the probe during this 16-month period, all the remaining fly-by images would have been lost forever.

And to put the distance into perspective, when a single data packet was sent from Pluto, it took over five hours for it to arrive to Earth, even though the transmission was occurring at the speed of light, 300,000 kilometers per second.

In another example literally closer to home, the common, low-end home Wi-Fi network has a theoretical maximum speed of 54 Mbps. Hence, in optimal conditions, the amount of data sent by *New Horizons* would pass through this kind of Wi-Fi connection in roughly 16 minutes instead of 16 months.

Regarding *cellular data networks*, the current top of the range is called *LTE Advanced* (where LTE is shorthand for *Long Term Evolution*), offering a theoretical speed of 300 Mbps. This would be able to deliver the *New Horizons* images in just under three minutes.

Depending on the number of simultaneous users that are accessing the same cellular base station at the same time, the real data speed will be downgraded. Similarly, your distance from the base station and any obstacles between your device and the base station play a big role here, but in practice it is possible to get several tens of Mbps of average download speed, which is very good for most purposes.

For city dwellers, having a fixed-line Internet connection via *fiber-optic data cable* is by far the best connection you can get, as commonly available urban connectivity speed is 100 Mbps, often both for uploading and downloading of data. And the available offerings are getting faster and cheaper all the time.

But whatever top-of-the line high-speed technology you have today will be old news in a couple of years, as new, more complex and hopefully more entertaining use cases will be invented. Our data communications solutions will keep on playing catch up against the ever-increasing needs: currently, *fifth-generation (5G)* networks are being rolled out, with a promise of a tenfold improvement over the prevalent 4G speeds, and hence, over a 5G channel, all of the *New Horizons* fly-by data would be transmitted in less than a second.

The other side of the speed coin is the fact that higher speeds mean higher power consumption: every data packet has to be processed, the potential encryption and compression need to be opened, and the data has to be stored or displayed somewhere in real-time.

With the increase in connection speeds, all this requires ever-increasing amounts of computing power, which in its turn has a direct relationship with the amount of energy needed for these operations. The advances in microchip technology do counteract this trend, as new *microprocessors* with smaller internal transistors can perform the same amount of processing with less power, but we humans have so far been extremely proficient in using all the available capacity as soon as it becomes available. Therefore, unless there is some fundamental breakthrough in battery technology, the available battery capacity will remain as the ultimate limiting factor to our wireless device usage.

# TechTalk
# There is No Free Lunch

You can't extract information from nothing.

In the world of radio, this rule of thumb means that even though your transmission is a pure *sine wave*, strictly on a specified *carrier frequency*, modulating the transmission with another signal in order to actually embed some information into your transmission will cause the transmitted signal to occupy a slice of adjoining frequencies instead of just a single, sharply specified one.

Your *channel frequency* will still be the one you are transmitting on, but your signal will "spill over" to adjoining frequencies on both sides of this frequency.

The resulting small block of adjacent frequencies is called the *channel*, and the *bandwidth* describes the size of the channel.

The simple consequence of this limitation is that the higher the maximum frequency of the modulating signal, the more bandwidth you need, and thus the wider your channel will be.

A simplified rule for analog transmissions is that your channel will be twice as wide as the highest frequency of your modulating signal.

Therefore, if you are transmitting on a frequency of 600 kHz and modulate your transmission with a telephone-quality voice signal that is limited to a maximum frequency of 4 kHz, your channel width will be 8 kHz, occupying frequencies from 596 to 604 kHz.

In order to avoid interference, adjacent transmissions must be separated at least by the bandwidth amount from each other. Therefore, if you have a fixed block of frequencies in use, the number of channels you can have in this block depends on how many times you can fit bandwidth-sized blocks, side by side, into the total available frequency block.

The frequency spectrum is divided into bands, and most of us are familiar with at least two common identifiers for this, as every radio tends to have selectors for AM and FM.

© Springer International Publishing AG, part of Springer Nature 2018
P. Launiainen, *A Brief History of Everything Wireless*,
https://doi.org/10.1007/978-3-319-78910-1

Although these abbreviations actually refer to types of modulation, as explained in Tech Talk *Sparks and Waves*, they also switch your receiver between two distinct frequency bands:

The AM band usually covers frequencies from about 500 to 1,700 kHz, which is officially known as the *Medium Frequency* part of the radio spectrum, and has been defined to contain channels of 10 kHz width in the Americas region. This means that the AM band can theoretically accommodate 120 individual channels of this size.

10 kHz bandwidth is good enough for talk radio but miserable for music, as the maximum modulating frequency has to be below 5 kHz. The frequency range for our ears is at best from 20 Hz to 20 kHz, so it is no wonder that the transmissions of any musical content on the AM band sound very low quality to us. But for talk radio use that's totally fine.

The benefit of using this low-frequency band for transmissions is that, thanks to the way these lower frequencies follow the curvature of the Earth, stations can be heard from great distances, as was discussed in TechTalk *Sparks and Waves*.

As the name says, all transmissions in this band are amplitude modulated.

Selecting "FM" on your radio usually offers frequencies from 87.5 to 108 MHz, which means that even the low range of this part of the radio spectrum has a frequency almost 50 times higher than the top part of the AM spectrum.

If you kept using the same channel bandwidth as for the AM band, this FM band could accommodate over 2,000 channels, but instead of maximizing the number of channels, the higher frequency is used to widen the individual channel to 100 kHz, thus allowing better quality audio transmissions: FM audio modulation bandwidth is 15 kHz, which is high enough for most of us with conventional hearing abilities, as the upper frequency range that our ears can hear tends to decay with age, and conventional music content usually has very little spectral information between 15 and 20 kHz.

The 100 kHz channel width theoretically allows just over 200 separate channels to be crammed between 87.5 and 108 MHz, and as the audio modulation does not use up all of the available bandwidth, there's ample room in an FM channel for other purposes than just sound. This extra bandwidth is often used to carry information regarding the station name and the song being played, so that this information can be displayed on the receiver, and as explained in Chapter 4: *The Golden Age of Wireless*, the wider channel also makes it possible to transmit stereophonic sound in a way that is compatible with monophonic FM receivers.

There are some practical limitations that halve the actual channel capacity of the FM band: one less understood one is related to the *Heterodyne Principle*, which is described in TechTalk *Sparks and Waves*, and which is used by all current FM receivers.

Due to the fact that the standard FM receivers use the super-heterodyne model, an intermediate frequency of 10.7 MHz is used for the local amplification of the received signal. This effectively makes every FM receiver act simultaneously as a low-power transmitter on a frequency that is 10.7 MHz higher than the current channel frequency.

Therefore, it would be very unwise to have two stations in the same coverage area with frequencies differing by 10.7 MHz: for example with one station at 88.0 MHz and another at 98.7 MHz, a nearby listener of the 88.0 MHz station would cause interference to another listener who has tuned her radio on 98.7 MHz, as the *local oscillator* of the receiver tuned in the 88.0 MHz channel would generate a low-power 98.7 MHz signal that might be strong enough to block out the reception of the 98.7 MHz transmission. The distance in which this might be a problem is at best calculated in tens of meters, but it could be a problem for car radio listeners in a traffic jam, or between neighbors in a block of flats.

This effect is easy to detect with two analog FM radios: tune one on some frequency and you will find a "quiet spot" on the dial of the other one at a location that is exactly 10.7 MHz higher.

Due to the use of this intermediate 10.7 MHz frequency, digitally tuned FM receivers in some countries only allow the selection of odd decimal frequencies, from 87.5 to 107.9 MHz, as the frequency of the local oscillator in this case will always end up on an even frequency decimal, on which there are no other receivers or transmitters. Every FM transmitter in the country will always transmit on a frequency that ends with an odd number. As a result, you cut the number of your potential channels by half, but all of these channels can be taken into interference-free use in high-density urban areas.

When your modulation signal becomes more complex, you end up generating a wider channel, and to accommodate multiple channels with highly complex modulation signals like television transmissions, you have to use even higher frequencies.

As an example, terrestrial television transmissions use the *Ultra High Frequency (UHF)* part of the spectrum, starting from 300 MHz.

The top of the range has microwave frequencies in the GHz range, where it is possible to cram several digital television transmissions into a single but very wide transmission channel, and as discussed in Chapter 7: ***Traffic Jam over the Equator***, these high-capacity channels are used for satellite communications, as well as for a multitude of point-to-point terrestrial microwave links.

As we enter the microwave portion of the electromagnetic spectrum, the absorption caused by water molecules starts affecting the reception of the signal. This phenomenon and its side effects were discussed in Chapter 11: ***Home Sweet Home***.

With microwaves, heavy rainfall or a snowstorm can momentarily reduce the signal strength of a satellite television broadcast or a low-power terrestrial microwave link to a point that reception is no longer possible.

Further up, when the frequency reaches a couple of hundred terahertz (THz) range, we reach the visible light part of the electromagnetic spectrum, in which we need only a thin physical obstruction to completely block out any transmission.

Increasing the frequency also increases the energy of the generated waves, in direct relation to the frequency. Therefore, when the frequency of the radiation goes past visible light into the ultraviolet part of the spectrum, the waves have enough energy to pass through obstructions again. Hard ultraviolet radiation, X-rays and especially gamma rays can penetrate deep into matter, and together they form the *ionizing radiation* part of the spectrum.

All three have enough energy to knock off electrons from the atoms that they hit along the way, and thus are harmful to living organisms. Ultraviolet radiation is strong enough to be used to kill bacteria, and it can also cause skin cancer as a result of excessive exposure to the incoming ultraviolet radiation from the Sun.

Detecting X-rays and gamma rays forms a major part of *Radio Astronomy*, allowing us to understand complex things like the formation of black holes and the processes that happen inside our own Sun. With gamma rays, we have reached the top end of the electromagnetic spectrum, with the strongest known type of radiation.

Another fundamental aspect of electromagnetic radiation is the concept of *wavelength*.

An electromagnetic signal travels through space as a *sinusoidal wave*, and the distance needed to complete one oscillation defines the wavelength of the signal.

Wavelength can be calculated by dividing the speed of light by the frequency, and hence the wavelength of a signal gets proportionally shorter as the frequency gets higher.

The wavelength varies from tens of thousands of kilometers for the *Extremely Low Frequency (ELF)* band (3–30 Hz) to meters in the television and FM radio bands (30–300 MHz), and down to nanometers (billionths of a meter) for the visible light spectrum (430–750 THz).

The wavelength of a signal affects the antenna design, as an antenna for a certain frequency works best if its length matches the wavelength of the signal in a fixed ratio. For example, a common *quarter-wave monopole antenna* design consists of a metallic rod that is one fourth of the wavelength of the desired frequency, combined with a suitable grounding surface.

Hence, as discussed in Chapter 3: **Radio at War**, the shift to higher frequencies made truly portable battlefield radios feasible, thanks to the shorter antennas they required, and our current gigahertz-band mobile phones can get away with fully internal antenna designs.

To summarize, the properties of electromagnetic waves differ wildly depending on the frequency band, and each frequency band has its optimal usage scenarios. The higher the frequency, the more modulation it can accommodate and thus more information can be transmitted per second, but the way different frequencies propagate in the Earth's ionosphere and the way they pass through intermediate matter like walls, rain and snow also varies hugely. Therefore, the selection of which frequency band to use depends totally on its planned application and the context it will be used for.

# TechTalk
# Making a Mesh

Digging cables into the ground or installing them on poles is very expensive, and hence has been a limiting factor for the improvement of communications infrastructure in many less-developed countries.

The value of the copper needed for the cables themselves has caused problems, as thieves have cut out the cabling and sold it onwards as scrap metal. Any copper cables are prone to failures due to thunderstorms, especially in rural areas where the cable runs are long and often exposed to weather: the longer the cable, the more subject it is to induction-induced over-voltages due to nearby thunderstorm activity: such induced peak currents can easily fry electronic equipment at the end of the connection.

Fiber-optic data cables do not have the same susceptibility to electric storms, but at the moment they require even more expensive infrastructure components, which often becomes a limiting factor, especially in rural settings.

Therefore, the possibility of switching directly to *cellular networks* has been a boon for many countries: building cellular towers at the heart of existing cities, towns and villages makes connectivity instantly available to all, without the cost and labor of connecting each individual household to a wired network.

You still need a complex and high-bandwidth wired or dedicated microwave-based connection between the *base stations* and the rest of the communications infrastructure, but you are only dealing with thousands, not millions, of individual connections.

I remember a visit to Kenya over ten years ago, during which I purchased a local SIM card for data usage and was astonished at the speed and quality that it provided in the outskirts of Mombasa. The cellular revolution has given many countries the ability to jump from zero to a totally connected society, thus giving a huge boost to local commerce. The positive economic effect of such cellular installations can be enormous.

And when the cellular phone is the only means of communicating, it is no surprise that cellular phone-based cashless payment systems, like the *M-Pesa* of

© Springer International Publishing AG, part of Springer Nature 2018    193
P. Launiainen, *A Brief History of Everything Wireless*,
https://doi.org/10.1007/978-3-319-78910-1

Kenya, had been around for almost a decade before starting to appear in the smartphones of the developed world.

The trend of "bypassing the cable" that has been so groundbreaking in the emerging market economies is making its way to the developed world, too: in the United States, the number of households with only cellular connectivity surpassed the number of households with fixed phone lines in 2017. These customers are not bypassing the cable due to the non-existent availability of such a connection, they are cutting themselves off of an existing one.

With "fast enough" fourth-generation networks, even the connection to the Internet is now becoming a wireless one, often subsidized by the telephone companies that are happy to see their costly wired infrastructure become obsolete.

But there are still numerous places where this kind of cellular technology based wireless revolution is not yet feasible: base stations are still relatively expensive to install and maintain, requiring a certain minimum return of investment to become feasible, and in some less-tranquil parts of the world, a continuous armed security presence must be deployed to ensure that the equipment at the base station, or the fuel stored for the resident backup power generator, is not stolen.

And what if your "customer base" is just a bunch of wild animals in the middle of a rain forest, hundreds of kilometers away from the nearest civilization, and your need for connectivity only occurs when one of your "customers" wanders in front of a hidden camera trap?

Or what if you want to provide a wireless Internet service that is going to be used by hundreds of occasional users in a tiny, far-flung village that has only one physical connection to the Internet?

You should build a *mesh network*.

The basic concept of a *mesh node* is radio-enabled equipment that dynamically adapts to its surroundings: it finds other compatible radios within range and maintains a connectivity and routing map of all other nodes in the network.

A good example of such a node would be a solar-powered wildlife camera with a radio transmitter/receiver that can cover a distance of a couple of kilometers.

You spread nodes like these in a suitable pattern, so that each one can connect to at least one other node. As each node further away connects to an additional node or nodes near to its own position, you can cover several hundreds of square kilometers with only a couple of tens of nodes and still be able to transmit data freely from one edge of the mesh to another: the message is delivered, hop-by-hop, by the adjoining nodes.

Connect one node of this mesh to the true Internet, and you can have connection from each node to anywhere in the world, and vice versa.

In populated areas you can use cheap Wi-Fi hardware with improved external antennas to have a mesh grid size of 200–500 meters, benefiting from the high bandwidth of Wi-Fi.

The drawback of a deep mesh network is the fact that the same data has to be re-transmitted several times until it reaches its destination, and therefore if the amount of traffic is high, your closest node is constantly using bandwidth and energy to relay messages from other nodes. Also, for cost and power management reasons, most solutions rely on cheap, standard hardware and have only one set of radio circuitry. Therefore, they can't send and receive at the same time, which effectively cuts the available bandwidth in half.

On purpose-built mesh hardware, it is possible to create separate radio channels for the core routing traffic and the network endpoint traffic, which would speed up the overall network throughput performance, but also increase the power consumption of nodes.

When a mesh like this is connected to the Internet, the bandwidth available for an individual user goes down rapidly if many others are trying to access the Internet at the same time, but in many cases, even a slow connection is much better than no connection at all. Normal browsing of web pages on the Internet also does not require a constant stream of data, so multiple users can easily be served, as their requests are interleaved in time. Just hope that not too many users are trying to watch YouTube videos or other streaming content at the same time.

In the optimal case, the network topology is planned to be dense enough so that the overall mesh is resilient against changes—if one node disappears for one reason or another, the routing can work around this and use other available nodes instead.

Therefore, the common denominator for all mesh networks is the automatic adaptability to the potentially hostile and constantly changing environment. In this sense, they mimic the functionality of the existing *Internet Protocol (IP)* networks, which also route each individual data packet according to the constantly changing routing situation of the Internet.

Apart from this basic requirement, mesh systems can be designed freely, depending on their expected usage model and available power limitations.

A good example of real-life implementation is the *FabFi* mesh, which is used to cover towns in Afghanistan and Kenya. The achievable throughput is over 10 Mbps, which is very respectable for occasional connectivity needs.

Mesh networks are not only a viable solution for developed countries: the *Detroit Community Technology Project* is maintaining a set of interconnected mesh networks as part of their *Equitable Internet Initiative* to alleviate the fact that Detroit is one of the least connected cities in the United States: 40% of the residents of Detroit have no Internet access. This kind of digital divide is becoming more and more of a cause for further alienation from society, as most services are increasingly only accessible via the Internet. Providing connectivity through a shared setup enables lowering the cost to a level that can support a large number of non-paying participants and in Detroit's case, help in the recovery of a city that has been hit by a major economic downturn.

The nodes in a mesh can even be portable handsets.

Many countries have deployed emergency radio networks based on *Terrestrial Trunked Radio (TETRA)*, in which the network is supported by a relatively low number of traditional base stations, but if some or all of those get knocked out due to a hurricane or an earthquake, the handsets can configure themselves to act as relaying nodes.

Naturally, if the relay usage is very high, the handsets are on all the time, so their battery life is greatly reduced, but in the case of TETRA, the slimness of the handset is not a driving design parameter, so they can be provided with much beefier batteries than your usual smartphones.

TETRA has its origins in the 1990s, and thus is voice-oriented and very poor in terms of data connectivity, but with its robust handsets and inbuilt mesh feature, it is hard to replace until handset-to-handset mesh becomes a viable feature in revised 4G or 5G specifications.

Mesh technology is one of the potential ways to support the new concept of the *Internet of Things (IoT)*, that was briefly discussed in Chapter 11: ***Home Sweet Home***.

With IoT, your environment is filled with simple sensors and other useful devices that are intermittently generating data and relay their information to some centralized processing unit when the need arises.

This "master node" then offers external Internet connectivity for the control and data access of the entire IoT mesh.

There are also proprietary military versions of mesh networking, which is understandable, as the battlefield environment is probably the best example of a highly dynamic situation in which reliable and resilient communications is of vital importance.

Mesh networks fulfill a "sweet spot" in cases where the expected traffic density and the distance between nodes happens to be well suited to the power capabilities of the nodes. With their dynamic adaptivity, they are very robust against changes in network topology, can be set up and expanded easily and offer yet another alternative for harnessing the power of electromagnetic waves.

# TechTalk
# The Holy Grail

For the actual radio spectrum, the different behaviors of these highly varying frequencies have been discussed in the other TechTalks above.

Similarly, problems encountered in creating these waves have been addressed along the way, from spark-gaps to generators to solid-state electronics, first with vacuum tubes and now with transistors and microchips.

The third important aspect of signal generation is the modulation that is either suitable for the frequency band in use, or optimal to the kind of information that needs to be transmitted, whether it is digital television or narrowband audio for mobile conversation. Again, this was discussed in detail in TechTalk *There is No Free Lunch*.

Even the best solid-state electronics have physical limitations in terms of amplifying low-level, high-frequency signals that are extracted from the receiving antenna. Therefore, it has been mandatory to turn this weak signal first into a lower-frequency one before further amplification and demodulation in the receiver. This application of the *Heterodyne Principle* was described in TechTalk *Sparks and Waves*.

Another limitation is the required antenna length, which usually can only be optimally tuned on just a small portion of the frequency spectrum.

All these constraints have, until very recently, forced the use of 100% purpose-built electronics, in which the utilized frequency band and supported modulation scheme are hardwired into the actual circuitry, and thus are practically impossible to change after the circuitry has been assembled.

Four recent technological advances are now challenging this mainstream approach:

First, it is now possible to create special transistors that have very low internal *stray capacitance*, allowing amplification of frequencies in the gigahertz range. This makes it possible to capture the faint signal from the receiving antenna and amplify it directly as-is, without any intermediate frequency conversion steps.

© Springer International Publishing AG, part of Springer Nature 2018    197
P. Launiainen, *A Brief History of Everything Wireless*,
https://doi.org/10.1007/978-3-319-78910-1

Second, digital signal processing is getting cheaper and faster, often thanks to the ever-increasing demands of our beloved digital "timewasters", like game consoles and High Definition (HD) digital television receivers.

Third, basic computing capabilities are still expanding along the original *Moore's Law*, which states that the number of transistors packed into the same physical space in microchips doubles approximately every two years. More transistors per area means that they become smaller, which in most cases reduces the internal stray capacitances and thus improves the maximum achievable switching speed and lowers the overall power consumption. This increase in processing speed can be directly utilized without any modifications to existing software.

We are also in the verge of moving from two-dimensional chip designs to three-dimensional ones, which allows a yet larger number of transistors to be crammed into our silicon chips: the more transistors at your disposal, the more complicated processing methods, like parallel processing, can be utilized.

Last but not least, smart, adaptive antennas are being developed, making it possible to transmit and receive on a much wider set of frequencies than with the traditional, physically fixed antennas.

Put all these together, and we are getting closer to the ultimate Holy Grail of wireless: *Software Defined Radio (SDR)*.

In an SDR receiver, the signal from the antenna is directly amplified, fed into a high-speed analog-to-digital converter, and then handed over to the signal processor circuitry, guided by traditional computer logic. The received signal is therefore turned directly into a stream of bits, the manipulation of which from that point forward is only limited by the available computing capacity.

Want to use the circuitry to listen to FM radio?

Just load the appropriate software.

Want to add a new modulation type to your existing mobile handset?

Just load new software that contains the necessary logic for handling the new modulation.

Want to send a search-and-rescue signal to a satellite from your mobile phone when you get stranded on a remote island without cellular coverage?

Just select the appropriate item from your menu, and the corresponding software module that emulates a standard *406 MHz Personal Locator Beacon (PLB)* is activated.

With fast enough SDR technology, combined with smart antenna solutions, your device would not become obsolete as new modulation technologies and frequency bands are taken into use. When a novel modulation method that improves the wireless data connection speed is invented, or a new frequency band that adds channels to your cellular connection is released for use, all you need is a software upgrade.

And instead of having different circuitry to handle your various connectivity modes from Bluetooth to Wi-Fi to cellular to GPS, you could run them all in parallel on a single, fast enough circuitry. The microchips that form the basis of the radio circuitry of today's devices already handle many of these different types of radios in parallel, thanks to the hardwired logic that is built in, but SDR would add an unprecedented potential for flexibility and future extensions.

Having fully programmable hardware would also speed up the development of new protocols, as anything could be tested in real life by only writing new controlling software: there would be no need to wait for the new circuitry to be first defined, then made into astronomically expensive first-generation microchips, only to be scrapped due to some minor problem that could not be foreseen in the process.

And when everything is based on software and there are clever programmers around, there's always the potential for misuse—an open SDR environment could for example be made to run rogue code that takes down entire cellular networks: the existing cellular networks are based on commonly agreed upon standards and hence expect correctly behaving equipment at both ends of the connection. Sending malformed data at the right time could easily take such a network down, and devising a program for an SDR for this purpose could be fairly easy.

Despite potential problems, we are rapidly heading in this direction. Hobbyist SDR boards can be bought for a couple of hundred dollars, and 100% software-based demonstrators of various protocols have been created with these devices, including the functionality of a GSM base station.

These do still have many practical limitations, especially in the width of the supported radio spectrum and the smart antenna technology, but the direction is clear—the moment SDR circuitry becomes cheap enough, it will become the heart of any wireless device.

This will be the Holy Grail that adds unprecedented flexibility to our exploitation of the electromagnetic spectrum.

# Index

© Springer International Publishing AG, part of Springer Nature 2018
P. Launiainen, *A Brief History of Everything Wireless*,
https://doi.org/10.1007/978-3-319-78910-1

Printed in the United States
By Bookmasters